Niklaus Fankhauser

Comparative Genomics of Membrane Proteins

Niklaus Fankhauser

Comparative Genomics of Membrane Proteins

in Tropical Parasites and their Hosts

Südwestdeutscher Verlag für Hochschulschriften

Impressum/Imprint (nur für Deutschland/ only for Germany)
Bibliografische Information der Deutschen Nationalbibliothek: Die Deutsche Nationalbibliothek verzeichnet diese Publikation in der Deutschen Nationalbibliografie; detaillierte bibliografische Daten sind im Internet über http://dnb.d-nb.de abrufbar.
Alle in diesem Buch genannten Marken und Produktnamen unterliegen warenzeichen-, marken- oder patentrechtlichem Schutz bzw. sind Warenzeichen oder eingetragene Warenzeichen der jeweiligen Inhaber. Die Wiedergabe von Marken, Produktnamen, Gebrauchsnamen, Handelsnamen, Warenbezeichnungen u.s.w. in diesem Werk berechtigt auch ohne besondere Kennzeichnung nicht zu der Annahme, dass solche Namen im Sinne der Warenzeichen- und Markenschutzgesetzgebung als frei zu betrachten wären und daher von jedermann benutzt werden dürften.

Verlag: Südwestdeutscher Verlag für Hochschulschriften Aktiengesellschaft & Co. KG
Dudweiler Landstr. 99, 66123 Saarbrücken, Deutschland
Telefon +49 681 37 20 271-1, Telefax +49 681 37 20 271-0, Email: info@svh-verlag.de
Zugl.: Bern, Universität, Diss., 2007

Herstellung in Deutschland:
Schaltungsdienst Lange o.H.G., Berlin
Books on Demand GmbH, Norderstedt
Reha GmbH, Saarbrücken
Amazon Distribution GmbH, Leipzig
ISBN: 978-3-8381-0566-6

Imprint (only for USA, GB)
Bibliographic information published by the Deutsche Nationalbibliothek: The Deutsche Nationalbibliothek lists this publication in the Deutsche Nationalbibliografie; detailed bibliographic data are available in the Internet at http://dnb.d-nb.de.
Any brand names and product names mentioned in this book are subject to trademark, brand or patent protection and are trademarks or registered trademarks of their respective holders. The use of brand names, product names, common names, trade names, product descriptions etc. even without a particular marking in this works is in no way to be construed to mean that such names may be regarded as unrestricted in respect of trademark and brand protection legislation and could thus be used by anyone.

Publisher:
Südwestdeutscher Verlag für Hochschulschriften Aktiengesellschaft & Co. KG
Dudweiler Landstr. 99, 66123 Saarbrücken, Germany
Phone +49 681 37 20 271-1, Fax +49 681 37 20 271-0, Email: info@svh-verlag.de

Copyright © 2009 by the author and Südwestdeutscher Verlag für Hochschulschriften Aktiengesellschaft & Co. KG and licensors
All rights reserved. Saarbrücken 2009

Printed in the U.S.A.
Printed in the U.K. by (see last page)
ISBN: 978-3-8381-0566-6

Contents

1	**Acknowledgments**	**3**
2	**Introduction: Biological Sequence Analysis**	**5**
	2.1 Sequence Data	7
	2.2 Genes	8
	2.3 Alignments	9
	2.4 Repeats	11
	2.5 Hidden Markov Models	12
	2.6 Neural Networks	14
	2.7 Clustering	16
3	**Identification of GPI anchor attachment signals by a Kohonen self-organizing map**	**23**
	3.1 Introduction	25
	3.2 System and Methods	26
	3.3 Algorithms	26
	3.4 Implementation	27
	3.5 Discussion	29
4	**Surface antigens and potential virulence factors from parasites detected by comparative genomics of perfect amino acid repeats**	**33**
	4.1 Background	35
	4.2 Results and Disussion	36
	4.3 Conclusions	41
	4.4 Methods	42
5	**Comparative transportomics between parasites and free-living eukaryotes**	**49**
	5.1 Background	51
	5.2 Methods	51
	5.3 Results and Discussion	53
	5.4 Conclusions	57
6	**List of Programs**	**59**
	6.1 GPI-SOM	61
	6.2 Reptile	61
	6.3 Dora	61
	6.4 Org-Get	62
	6.5 circDend	62
	6.6 memProtPlot	62
	6.7 ReduceToMax	63
	6.8 KohoNet-GUI	63
	6.9 PfamOrg	63
	6.10 paraFree	63
	6.11 protCluster	64
	6.12 TransOrgalin	64
7	**Abbreviations**	**67**

CONTENTS

Chapter 1

Acknowledgments

I would like to express my gratitude to Prof. Pascal Mäser, whose inspiring supervision made this work possible. I also wish to thank Prof. Erik van Nimwegen for evaluating my thesis as an external co-referee and Prof. Thomas Seebeck for being my co-examinator as well as for good advice.
For their immensely helpful support of my works I feel grateful towards Prof. Isabel Roditi, Prof. Ernst Schweingruber, Prof. Peter Bütikofer, Prof. Beatrice Lanzrein, Prof. Walter Senn, Dr. Joël Adler and Dr. Paul Dupree.
I thank all current and past members of the Mäser group for providing a inspiring work environment. This work would have been impossible without the help of the Informatikdienste of the University of Bern for resources and support.
This work was financially supported by the Swiss National Science Foundation and the Roche Research Foundation.
I thank my family and all my friends for supporting me and giving helpful advice.
Special thanks to Gabriela Marti.

Chapter 2

Introduction: Biological Sequence Analysis

The aim of my PhD project was to take advantage of the available genomic sequence data to perform comparative genomics between free-living and parasitic organisms regarding predicted membrane proteins.

As first discovered in parasites[1], not all membrane proteins are embedded into the membrane, some are anchored to it by lipid molecules like for example GPI-anchors. Transmembrane prediction programs easily confuse hydrophobic signal sequences with transmembrane domains, leading to false positives. The GPI-anchor is only attached to proteins containing such signal sequences. Therefore, increasing the efficiency of predicting signal sequences also improved the prediction of transmembrane proteins, because fewer false positives are found. More about the in silico detection of GPI anchors in Chapter 4.

Another feature often encountered in proteins located at the surface of cells are internal repeats. Finding an accurate quantitative measure of such repetitive proteins was also an important part of this project and is described in Chapter 5.

Membrane proteins are often organized in families, so we also used methods to cluster proteins into families. This process is explained in Chapter 6.

The Introduction gives an overview of the computational methods used from a molecular-biological point of view. Much could be told of the intricate techniques used to implement these methods and how to most efficiently process such large amounts of data using modern microprocessors, but this is beyond the scope of this Introduction.

1 Sequence Data

The fact that all life is build of cells has been known for more than 400 years[2], but many things about their workings are still unknown. The reactions required to salvage and use energy are catalysed by proteins, which are created by using the information stored in the DNA making up the genomes of cells. In order to fully understand the inner workings of a cell, the function of each of its proteins as well as their interactions and regulations has to be known. Many proteins have been extensively studied and characterised. But as the process can take years for one protein and even comparably simple cells have thousands of proteins, this could be enough work to occupy generations of researchers. There is hope that it could be completed much faster by using the information stored in DNA.

1.1 Sequencing

By using the chain-termination method[3], the nucleotide sequence of DNA can be determined automatically. This is made possible by the simplicity and elegance of this sequencing method, as compared to protein analysis, where each molecule is a singular challenge on it own. Regions of purified DNA are replicated by temperature cycled polymerase (PCR), using a fraction of labelled chain-terminating nucleotides, which are incorporated into the newly synthesised DNA strand by the polymerase, but do not allow it to attach any subsequent nucleotides. This chain-termination produces all possible fragments of lengths according to the sequence position of the nucleotide. They are distinguishable by the specific fluorescence wavelength of each dideoxynucleotide label. The fragment lengths are detected by capillary electrophoresis and the resulting chromatograms contains the DNA sequence. These steps can be carried out automatically for hundreds of DNA segments in parallel in very short time, making it possible to sequences genomes as large as that of the human with a total of approximately 3 billion[4] DNA base pairs.

1.2 Whole genome shotgun method

A limitation of the chain-termination method is that only short stretches of a few hundred base pairs can be sequenced in each reaction. Sequencing whole genomes is made possible by shredding them into small fragments[5], the molecular analogue to the effect of a shotgun. Multiple copies of the genome are fragmented to get randomly different cutting positions. When all segments have been sequenced, the genome can be assembled by a computer program using the following hashing method: For each fragment, it either just stores the sequence in memory or extends a stored fragment if there is a region of significant overlap. Fragments amounting to 8-10 times the size of the human genome had to be sequenced to get 99% of the correct data[4]. Chromosome maps created by genetic analysis provide anchor points to correlate sequences to the physical chromosome structure.

1.3 New methods

It is desirable to increase the speed and accuracy of the sequencing process as well as making it possible to sequence longer fragments in one run. A method called 454 sequencing[6] uses massively parallel sequencing by synthesis on a solid support. This means that a whole genome could theoretically be sequenced in one run by carrying out the reactions on nano-particles instead of separate tubes. The first step is again shearing the genome, then ligating adapters to the fragments and binding them to beads captured in droplets of an oil-emulsion PCR reaction. Amplification in each droplet results in each bead carrying 10 million copies of unique DNA template. Finally, pyrosequencing[7] detects nucleotide incorporation by the release of inorganic pyrophosphate and the generation of photons.

Polony sequencing[8] is similar but it uses hybridisation instead of replication to sequence the amplified segments.

While these methods have the advantage of using parallel sequencing, they still have short read lengths of only about 50 base-pairs. But it should be possible to improve this by further optimizing the procedure. A different approach is using mass spectrometry to either identify the fragments created by the chain-termination method or to sequence DNA fragment by MS-MS[9]. While the first possibility has advantages over gel electrophoresis based systems, the latter still has too short read lengths. Another new possibility is first converting DNA into RNA, because the resolution and detection ability of MALDI-TOF MS is increased for RNA[9].

2 Genes

Although genomes are interesting on their own, they only become useful to studying proteins when the positions of the genes encoding them have been mapped. While most bacterial genes can be found by searching the Shine-Dalgarno[10] ribosomal binding sequence, there is no universal pattern identifying genes in all organisms. Because the translation system of eukaryotes is more complex due to additional regulation systems, computer programs predicting genes have to consider many factors.

The three most successful programs, GlimmerHMM[11], Genescan[12] and Genie[13] use a statistical approach called a Hidden Markov Model (HMM)[14] to find genes. This approach, explained in Section 5, combines existing knowledge about genes with optimal parameters determined by training on experimentally known genes. The following two observations illustrate what kind of existing knowledge helps to find genes: First, they have to be in an open reading frame, spanning a sufficient long sequence between an initiation (ATG) and termination (UGA, UAG, UAA) codon. They have a tendency to start in regions of high CpG dinucleotide[15] abundance and promotor regions usually show a greater degree of sequence conservation when compared to genomes of related organisms. If biological information like this is incorporated in prediction programs, their accuracy can be increased.

2.1 Databases

Gene finding programs give a rough indication of where genes might be located, but their accuracy in finding the exact boundaries of genes is rarely above fifty percent[11]. This makes hand curation of genomic sequences necessary. The resulting lists of genes are stored in huge databases like Genbank[16] and Expasy[17], where each gene should theoretically be accompanied by everything known about its function as well as how certain these finding are. These databases are of great importance, because they provide a common set of genes for further research and even tough not all genes are certain, the probability of that certainty is available.

The quality of gene prediction can further be improved by taking into account other data sources beside the genome of an organism. One possibility is the comparison to genomes of other organisms, which makes it possible to quantify the amount of evolutionary conservation of a stretch of DNA. While DNA segments not encoding any genes are important on their own, mutations in segments containing genes have a much higher chance of making an organism unable to survive. Therefore, genes are frequently located in conserved regions of the genome. This information can be used to improve gene predictions. The benefit from this approach increases with each additional organism sequenced, so it is desirable to sequence all know organism.

An other way to improve gene predictions comes from expressed sequence tag (EST)[18] databases, which are produced by sequencing the beginnings of cloned cDNA molecules. The resulting short sequences may contain some errors because the fragments have only been sequenced once, but they represent genes that have been transcribed in the studied cell. Therefore they have higher probability of being translated to protein than a sequence not found in the EST database. By using alignments, described in Section 3.1, the ESTs can be matched against genomes despite the sequencing errors they might contain, so that new genes can thus be found.

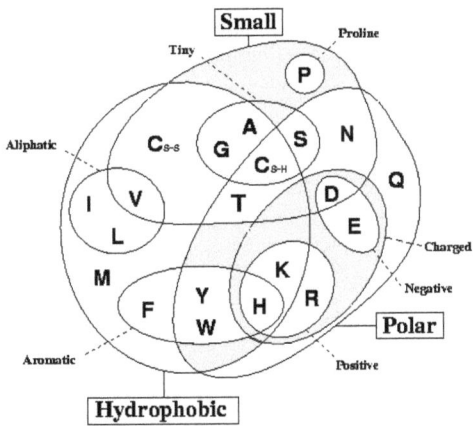

Figure 1: A Venn diagram grouping amino acids according to their properties, adapted from Livingstone & Barton[19].

3 Alignments

An alignment is a method to find similar DNA or amino acid sequences using a query sequence. Sequence similarities are of importance because DNA sequences accumulate mutations during evolution. The selectional pressure on retaining the function of a protein keeps most mutations from changing or abolishing its function. This means that alignments can be used to derive the function of a protein of previously unknown function by finding sequence similarities to others with experimentally verified functions. Base pair changes are not the only possible kind of mutations, there can also be insertions and deletions of one to whole domains of base pairs. These are called gaps in an alignment.

If a program would simply check the quality of all possible gaped alignments, this would take an extremely long time to compute. This made it necessary to use more efficient algorithms.

There also has to be a quantitative measure of how similar one nucleotide or amino acid is to another. For nucleotides, there are not many possibilities and they can be derived from the frequencies of the various possible mutation events.

Looking for common physicochemical properties of amino acids, as pictured in Figure 1, could help to get a measure of their similarities. But a more objective way is to align sequences of proteins with known functions from different organisms, so that the evolutionary conservation of each amino acids can be determined statistically. This substitution table is called a similarity matrix. Figure 2 shows the commonly used matrix is BLOSUM62[20], the BLOck SUbstitution Matrix created from sequences

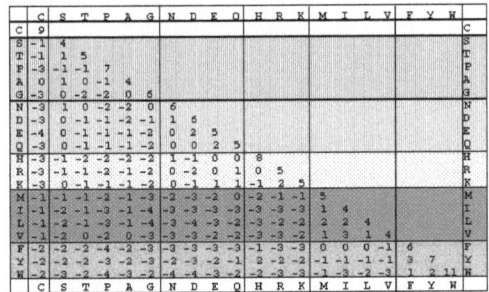

Figure 2: The BLOSUM62 similarity matrix. It contains the scores for all possible amino acid substitutions.

of at least 62% identity.

3.1 Global alignment

A global alignment has to find the optimal pairing of two complete sequences, in contrast to a local alignments, where the aim is to find the optimal alignment of parts of these sequences. Instead of considering all possible ways to align two sequences, the Needleman-Wunsch algorithm[21] optimizes this process by a dynamic programming technique.

It works by breaking down the problem into subproblems recursively, until it becomes possible to solve one. The way to break the problem into subproblems is by looking for the optimal way to align sub-sequences of the whole alignment. A matrix with one dimension the length of the first sequence and the other the length of the second sequence can store the results of all optimal sub-sequence alignments. Figure 3 provides an example of such a matrix.

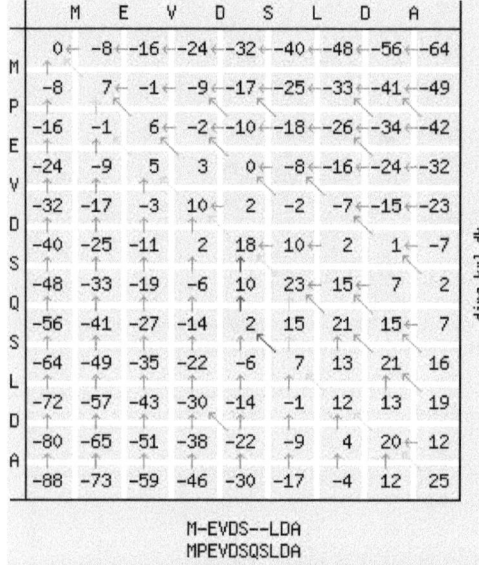

Figure 3: The alignment matrix created by the Needleman-Wunsch alignment of the sequences MPEVDSQSLDA and MEVDSLDA. The trace-back is shown graphically. Green arrows correspond to the optimal alignment shown; blue arrows correspond to alternative optimal alignments; and red arrows correspond to the possible trace-back directions for every given matrix cell. Image created by a script by Peter Sestoft, available at http://www.itu.dk/people/sestoft/bsa/graphalign.html

The first row and column of the matrix are filled by adding up the negative gap penalties, because these cells represent scores for no match. The first solvable sub-problem is the optimal alignment of only one amino acid (or nucleotide). In order to find the optimum, three possibilities have to be considered. The first is to align the two letters to get their similarity score added to the score from the previous cell (up and left to it). The second is to introduce a gap in the first sequence and get a gap penalty added to score from the cell above. The last possibility is to introduce a gap in the second sequence and get a gap penalty added to score from the cell to the left. This highest scoring of these three possibilities is now stored in the cell at the upper left corresponding to the start of both sequences. Now it is possible to calculate the next three adjacent cells of the matrix in the same way. This process is repeated until the whole matrix is filled. The score in the lower right field of the matrix is the score of the optimal alignment, because solving this last sub-problem is not actually a sub-problem anymore but the whole problem, as the sub-sequences now reached their maximal length. For each cell of the matrix a pointer to the cell from where the maximal score originated is saved. By following these pointers backwards to the beginning, the alignment of the sequences can be constructed.

3.2 Local alignment

The aim of local alignments is finding regions sharing close sequence similarity in distantly related sequences that cannot be aligned as a whole. The Smith-Waterman algorithm[22] for finding local alignments is a variation of the Needleman-Wunsch method. The only difference is that for finding the optimum score of sub-sequence alignments, it is possible to set a cell value to zero if all other possibilities are negative. This corresponds to starting a new alignment. Because of this change, the method for constructing the alignment from the matrix has to be slightly modified. The score of the optimal alignment is not in the lower right corner of the matrix anymore, but can be anywhere in the matrix. It can be found by looking for the cell with the highest score in the matrix. From this cell, the alignment can be constructed backwards as explained above with the only difference that it stops when a cell value is zero, because this is where the alignment ends. By starting the trace-back at the cell containing the second highest value (and below), sub-optimal alignments can be considered because they may also have biological significance.

3.3 BLAST

The above methods solve the problem of aligning two sequences optimally, but the running time is proportional to the product of the sequence lengths of the sequences to be aligned; their complexity is $O(n^2)$. This property makes it impossible the get results quickly. Performing on-line database searches would take hours if not days to finish with the above described methods.
The frequently used Basic Local Alignment Search Tool (BLAST)[23] uses heuristic methods to make this

process more than 50 times faster. Using a heuristic algorithm means that is no longer possible to prove that the solution is correct. It makes use of the sensible assumption that most good alignments contain some part of identity. Searching these identities takes very little time and allows the algorithm to save time later by not having to fill all of the matrix cells. It starts at matrix cells corresponding to identities (length 3 for amino acids) and extends them until the local alignment ends. This method might theoretically miss some optimal alignments, if they contain not enough identity. But in practice this is rarely the case and this method is widely used for database searches.

3.4 Multiple alignments

If more than two sequences need to be aligned, the process is called a multiple alignment. It is used to analyse protein families, which can have members in different organisms (orthologs) or in the same organism (paralogs). The multiple alignment of such sequences shows which regions or domains are conserved in a protein family. Studying these conserved regions in the laboratory is a frequently used approach to find the common biological function of the family. A multiple alignment can also visualize the amount of sequence divergence accumulated by mutations in different species and give insight into how they are related.

The main difficulty in creating a multiple alignment is the computational complexity of the problem. Theoretically, it would be possible to create them by extending the Needleman-Wunsch algorithm to multiple sequences. For three sequences, the alignment matrix shown in Figure 3 becomes a three dimensional cube. For increasing numbers of sequences, the number of matrix cells becomes impractically large, as this algorithm has an exponential complexity $O(c^n)$. It is not feasible to compute the multiple alignment of more than eight sequences[24] by this method. But protein families can have hundreds of members even in a single organism. Therefore, a more efficient way of solving this problem is needed. The most frequently used methods to solve this problem uses a heuristic approach.

The first step is to compute an alignment score matrix by performing Needleman-Wunsch alignment of all sequences against each others. It is also possible to use a faster method like BLAST in this step, because these alignments are not part of the final multiple alignment. They are only used to create a phylogenetic guide-tree of the sequences to be aligned. This tree will be used to create the multiple alignment in the next step. The tree is constructed by using the Neighbour-Joining method[25] described in Section 7. Now the alignment can be created progressively by starting from the two most closely related sequences, representing the two leaves of the tree with the shortest distance in between. Following the branches of the tree from the leaves to the root, one sequence after the other is now added to the alignment. The previously aligned sequences are unmodified and considered as one when aligning an additional sequence. Position-specific gap parameters are derived from the alignment of the first two sequences and used for the following alignments. After adding the last sequence, the multiple alignment is complete. This method is used by the ClustalW program[24], which is the most mature[26] and popular program for creating multiple alignments. It produces very reliable multiple alignments, except when used on sequences that are only distantly related.

Iterative methods can be used to improve the quality even in this case. They work like the progressive method described above, but in each iteration of adding a new sequence to the multiple alignment, the previous alignments are adapted using the more accurate gap parameters. The substitution matrix is also adapted in each iteration to more accurately match the evolutionary distance as determined by the phylogenetic tree. This method is used by the relatively new program Muscle[27], which quickly creates alignments of high quality[28].

4 Repeats

Repeats in amino acids sequences are of interest for a variety of reasons. It has been shown that such repetitive sequences are frequently created by the replication machinery of cells[29] and that it is possible to create proteins of new function by repeating existing sequences instead of going from a protein of one function to one with another by point mutations.

Repeats are used in proteins that have to bind to other regular structures like ice crystals in the case of ice-binding proteins[30]. Titin, the molecular ruler determining the length of muscle fibres, is on other example of a large protein structure composed of repeats[31]. Repeats are also found in large structural proteins such as collagen[32], the main protein in connective tissue of animals. Protein-protein interactions are frequently mediated by ankyrin repeats[33], a very common structural motif of 33 amino acids length. DNA binding proteins often use zinc finger domains[34] for attaching to specific nucleotide sequences.

4.1 Sub-optimal alignments

As described in Section 3.2, the score matrix of local alignments can be searched for sub-optimal alignments. If a sequence is aligned to itself, sub-optimal alignments can be used to reveal repeats[35]. Each of the sub-optimal alignments has a certain score, depending on the values for gap penalties and the substitution matrix. It has been shown that these scores obey Poisson statistics[36]. The distribution of scores derived from sub-optimal alignments of random sequences defines the parameters for the Poisson function. The Poisson function can now be used with these parameters to compute the significance from score and length of the sequence.

The length of the repeat unit can be found by creating a vector, where for each amino acid in the sequence the number of its occurrences in a sub-optimal alignment is noted. This is a projection of the path matrix into one dimension and it resembles a step function. By determining the size of the steps, the repeat length can be found. A drawback of this strategy is that for perfect repeats, it does not find the repeat boundaries as well as the hashing method described in the following Section. Examples of programs using the sub-optimal alignment method for finding repeats in proteins are Internal Repeat Finder[37], RADAR[38] and REPRO[39]. While RADAR provides an accurate detection of repeat boundaries, only the Internal Repeat Finder provides a p-value to estimate the statistical significance.

4.2 Word hashing

This approach finds perfect repeats only, but their boundaries are exact and the results can be used in comparative genomic analysis. We implemented this algorithm in the program Reptile[40] and a similar approach was used to create the RepSeq database[41].
In this method, all possible sub-strings (words) from length two to an arbitrary maximum of an amino acid sequence are generated. The number of occurrences of these words is counted and associated to the repeat sequences in a data structure called a hash table. After removing redundant repeats that are contained within longer ones, the repeated sequences are sorted by ascending P-value. This is a brute force method for generating and evaluating all possible repeats in a sequence. It has a very large complexity for a worst-case input sequence. But by optimisation and using realistic parameters for the word size, it runs below N^2 time for sequences of length N. The p-value is obtained by dividing the possible number of sequences containing repeats by the total number of possible sequences. The inclusion-exclusion principle corrects for repeats that have been counted too often (see Chapter 5). Finally, the p-value has to be adjusted to non equal amino acid distributions in a certain organism.

4.3 Fourier analysis

Multichannel Fourier transformation[42] is a method for detecting weak periodic patterns in the background noise of amino acid sequences. The result is a combined power spectrum, analogous to a p-value, which automatically extracts the most significant nonrandom features from the transforms of the original amino acids. Starting from the separated Fourier transforms of the 20 different amino acids, each amino acid represents one channel of information. The transforms are combined together, by assigning weights to them, in a way which separates the statistically significant features from the random noise. The combined power spectrum which emerges from this process is automatically scaled in terms of its significance.

This method was used to create the database of Tandem Repeats In Protein Sequences[43] (TRIPS). While it may detect very weak repetitive features, no measure of statistical significance of the repeats is provided. Indirect (aperiodic) repeats can also be detected using this method.

5 Hidden Markov Models

A Hidden Markov Model[44] (HMM) is an extension to Markov chains[45]. They are a statistical model of sequences, which do not have to be biological sequences; it could also be sequences of events for ex-

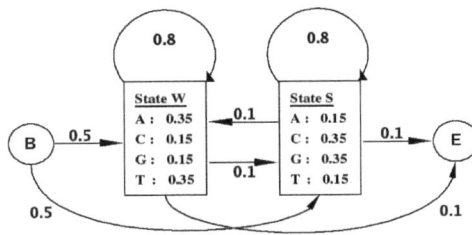

Figure 4: A hidden Markov model describing two hidden states in DNA sequences. The symbols B and E correspond to the beginning and the end of the sequence. The arrows represent the transition probabilities from one state to the other. Image taken from a tutorial at the Swiss Institute of Bioinformatics.

ample. They describe the probability of the state at the next positions at each position of the sequence. The state of each position is therefore only determined by the state of the position before it. For each state, the model defines what the transition probabilities are to get to any other state. The states can be nucleotides or amino acids. Markov chains are called like this because drawing the transition probabilities as arrows between the state symbols looks like a chain. But the information it contains could also be represented like a similarity matrix without diagonal symmetry. As illustrated in Figure 4, a HMM is an aggregation of multiple Markov chains, connected by transition probabilities for switching from one chain to the other.

5.1 Domain prediction

HMMs used for domain prediction can best be explained by picturing the example of predicting a functional property of a sequence, like transmembrane domains. In the simplest case, there is a Markov chain with transition probabilities for all amino acids in transmembrane domains, and another one for sequences outside of them. The model can be seen as a sequence generator, emitting amino acids with a certain probability depending on what Markov chain it is currently using. But from the sequence it generates, it cannot be seen which chain was used. This state of the model is, as the name HMM implies, hidden.

It is possible to find the most likely sequence of hidden states generating any given sequence using the Viterbi algorithm[46], if the transition probabilities of the model are known. The sequence of hidden states would correspond to the positions of the transmembrane domains in our example. A simple algorithm would find the most probable sequence of hidden states by listing all possible sequences of hidden states and finding the probability of the observed sequence for each of the combinations. This approach is theoretically possible, but to find the most probable sequence by calculating each combination is computationally too expensive. The Viterbi algorithm is a computationally efficient way of analysing observations of HMMs to find the most likely underlying state sequence. Recursion is used to reduce the computation time and the context of the entire sequence is used to quantify the noise. The algorithm calculates a partial probability for each sequence position, together with a back-pointer indicating how it could most probably be reached. The most likely final state is taken as correct, and the path to it traced back to the beginning via the back pointers. This is similar to how the Needleman-Wunsch alignment method described in Section 3.1 works.

The task of finding the structure and parameters of the HMM which best account for a large amount of data is called the training problem. The most difficult task is to adjust the model parameters to maximize the probability of the training sequences given the model. This is done by the Baum-Welch[47] algorithm, which iteratively improves the probability of a particular output sequence determined by the forward-backward algorithm until the parameters converge. This maximum likelihood estimation of the HMM is is very useful in practice. It is possible that this method does not find the best parameters, but this can be circumvented by carefully designing the model and using heuristics. The Baum-Welch algorithm is a variant of the expectation-maximization algorithm used in statistics to determine the unobserved variables in probabilistic models. It has a time and space complexity of $O(N^2 T)$, where N is the number of states in the model and and T is the length of the training sequences[48].

5.2 Protein family profiles

The HMMs used to describe protein families are using a similar concept as multiple alignments[49]. They contain a sequence of match states corresponding to positions in a protein or multiple alignment. Each of these states generates a letter from the amino acid alphabet according to certain, distinct distributions. There is also a delete state for each match state that produces no amino acid and skips that state. Insert states to either side of the match states generate amino acids in exactly the same way as the match states, but using separate probability distributions. There are three possible transitions from each state to another. Transitions into match or delete states always move forward in the model, whereas transitions into insert states do not. Multiple insertions between match states are possible since transition from the insert state to itself is allowed.

The parameters of the HMM are estimated from training sequences using the Baum-Welch algorithm, described in the previous Section. Now the HMM can be used to search for other members of a protein family of interest or sequences containing a given domain.

This method also produces multiple alignments of good quality in $O(n)$ time, as the number of states in the model is constant.

Evaluations using existing protein families showed that HMMs are able to distinguish members from non members with a high degree of accuracy. HMM profiles use a valid statistical treatment of insertions and deletions. Since in standard profiles, the optimal insertion/deletion scores can only be found by trial and error, the statistical significance has to be evaluated by empirical methods.

The first HMM-based profile methods required more than 100 sequences for good homologue recognition. By incorporating prior information about amino acid substitution probabilities into HMMs, effective HMMs can now be constructed from a handful of sequences.

6 Neural Networks

Neural networks have a dual biological significance; they are modelled after biological neurons and they have been successfully applied to biological sequence analysis. Inspired by the architecture of neurons in the central nervous system, a group of artificial neurons can be used as an adaptive system that adapts its structure to a specific problem during a learning phase. Neural Networks are used as models for the relationship between input sequences and their properties. They can also distinguish categories in data. In a neural network model, nodes are connected together to form a network. These are also called neurons, units or perceptrons. A perceptron, depicted in Figure 5(a), sums up a number of inputs with associated weight factors and passes the result to a (often sigmoid) transfer function. The practical use comes from algorithms designed to adapt the strength (weights) of connections in the network to approximate the desired output.

6.1 Feed-Forward Networks

This is the first and simplest type of artificial neural network. As shown in Figure 5(b), a feed-forward neural network is composed of multiple layers of neurons[50]. The information moves forward from the input nodes, through optional hidden nodes and to the output nodes. There is no connection from one perceptron to another in the same layer.

For each node, the sum of the products of the weights and the inputs is calculated. If the value is above some threshold the neuron fires and takes the activated value, otherwise it takes the deactivated value. Each neuron in one layer has connections to the neurons of the subsequent layer. A typical task for a neural network is to classify different inputs into categories. The learning process begins with presenting a pattern to the input layer. The pattern is transformed in its passage through the layers of the network until it reaches the output layer. Every unit in the output layer represents a different category. Beginning with randomly assigned weights, the outputs of the network are compared with the desired outputs, the correct classification. The unit with the correct category should have the largest output value and the output values of the other output units should be very small. Based on the difference of this observed with the desired output pattern, all the connection weights are adjusted a little bit. The next time this same pattern is presented at the inputs, the value of the output unit that corresponds with the correct category will be higher than it is now and all other output values lower. So the differences between the actual outputs and the correct outputs are propagated back from the input layer to lower layers to be used at these layers to modify connection weights. This is why it is also called a back-propagation network. The number of cycles required to train the network has to be determined empirically and depends on the number of inputs, outputs and hidden neurons. If too many learning cycles are used, the network becomes over-trained and is not able to correctly categorise patterns outside the train-

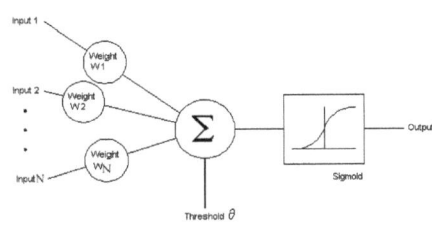

(a) Perceptron

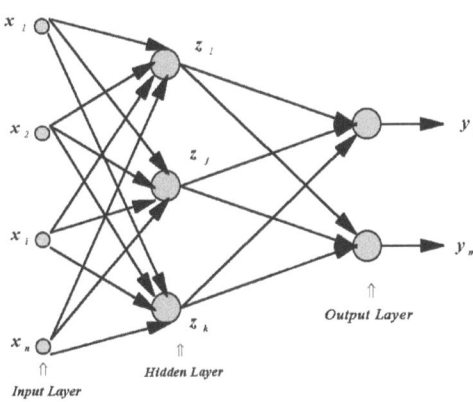

(b) Feed forward network

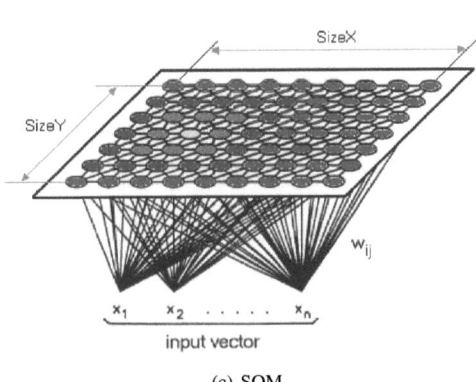

(c) SOM

Figure 5: (a), Schematic representation of a perceptron with weighted inputs and sigmoid transfer function by Nikolay Nikolaev; (b), Overview of the three levels of neurons of a feed forward network[51]; (c), diagram of a Kohonen self organizing map by Hans Lohninger.

ing data.

The most difficult step is how to represent the biological sequence data to the input neurons, because they need a numerical value. There are many possible ways to convert a DNA or amino acid sequence into a vector of numerical values, but it has to be adapted depending on what problem should be solved. The input vector can be composed of different parts describing specific properties of sequences. Useful properties are amino acids frequencies and distribution as well as chemical interpretations of amino acids or results of pattern matching.

Examples of programs using feed forward networks are PHD[52] and Secondary[53], programs for predicting secondary structure of protein sequences as well as Myristoylator[54], a tool to predict N-terminal myristoylation of proteins.

6.2 Self Organising Maps

Self-organizing maps (SOMs), also called Kohonen maps after their inventor, are a type of neural networks used for classification of hidden information in large data-sets[55]. As shown in Figure 5(c), they consist of an input layer of feed-forward network where the outputs arranged in a two or three dimensional grid. Each input is connected to all output neurons. Learning in SOMs also happens by adjusting the weights of the connections between units until the activated output neurons are well distributed on the map. In contrast to feed-forward nets, SOMs are performing unsupervised learning, guaranteeing minimal human bias. They distinguish patterns without knowing if and how many different patterns the input contains. If a two dimensional grid of neurons is used, the prediction of the SOM can be visualized as a two-dimensional map. On this map, similar input pattern excite neighbouring neurons. This is because the weights of a whole area are moved in the same direction in the training phase. Similar input samples appear mapped close together and dissimilar ones apart. The map can therefore give an estimate of certainty of the classification of a single input sequence. If the activated neuron is embedded in a portion of the map known only to be excited by one group of inputs, the certainty of the prediction is better than in the case where the activated neuron is on the border between a region of the map. Finding the best way to present the sequence data to the input neurons for a given problem is also a central problem

with SOMs.
Examples of successful applications of SOMs are the classification of DNA sequences based on nucleotide frequencies[56] or virtual potentials[57]. Analysis of protein sequences was carried out using bipeptide composition as inputs[58] and by GPI-SOM[59], our program for the prediction of GPI-anchored proteins (Chapter 4).

7 Clustering

Everything can be clustered, as long as there is some distance measure between each of the objects. The most common application is a tree of organisms that can be created if the evolutionary distance between each of them is known. Usually, a highly conserved sequence is used as a measure of distance. An example is the 16S rRNA, which is a part of the ribosome needed in all organisms for protein synthesis.

Global alignments of all possible sequence-pairs in a group of sequences provides a matrix of scores. Using these scores as a distance measure, various clustering methods can be used to create a tree.

A tree is called rooted if it is contains an outgroup. This can be an inferred most common ancestor sequence. Any sequence less similar to all in the tree than any two in it, but not by much, can be used as the root. See Figure 6(a) for an example of a rooted tree.

An unrooted tree shows the relations of the group of sequences without assuming a common ancestor, as shown in Figure 6(b).

It is also possible to cluster points in a space of arbitrary dimensions, using a distance measure like euclidean or city-block distance. Using these distances, the clustering methods described in the following sections can be applied. This is useful for analysing micro-array experiments, where each point can be a gene whose coordinates are defined by its expression levels under various conditions. We applied this methods to clustering organisms (Chapter 6) by treating them as points whose positions are specified by the number of hits for a number of HMM profiles matching important transporter families.

7.1 UPGMA

The most simple way to create a tree is by using UPGMA (Unweighted Pair Group Method with Arithmetic Mean)[60].

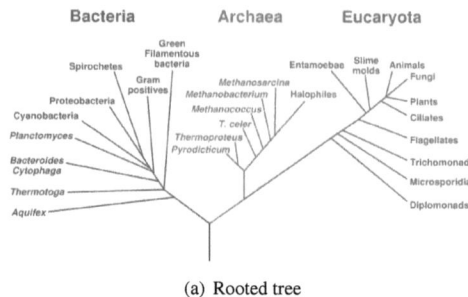

(a) Rooted tree

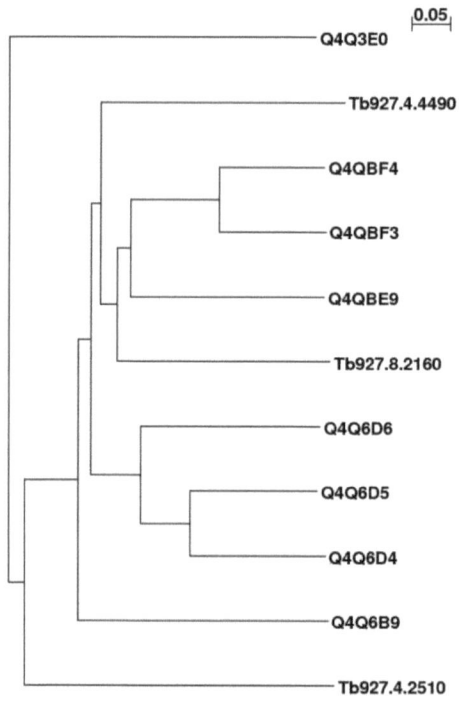

(b) Unrooted tree

Figure 6: (a), rooted tree of taxonomic kingdoms bases on rRNA genes using an inferred common ancestor sequence as outgroup; (b), unrooted tree of multidrug resistance proteins in both *Trypanosoma brucei* (Tb...) and *Leishmania major* (Q...). The tree was created by ClustalW using the Neighbor-joining method and displayed using NJplot.

It starts by clustering the two sequences most closely related according to the distance matrix. A new matrix is created where clustered sequences are treated as units whose distance to sequences is the arithmetic mean to the distances of the sequences it contains. The tree is now constructed by repeatedly clustering the two most closely related sequences or units until no sequence is left.

This methods is used by clustering software[61] by Michael Eisen, which we used to visualize the results of our genomic analysis of transmembrane proteins (Chapter 6).

7.2 Neighbor-joining

This method starts with the assumption that all sequences are equally related to each other[25]. This corresponds to a tree that looks like a sea urchin, showing no hierarchical structure. It is possible to derive the sum of the branch lengths of this simple tree from sum of the distances divided by the number of sequences.

Now the branch length of the tree is calculated assuming that two sequences are more closely related to each other, while the rest of the sequences still have equal relations. This is accomplished by finding the sum of the least-squares estimates[62] of branch lengths. By trying all possible pairs, the one with the shortest branch length can be found. In the next step, these two most closely related sequences are considered as one, one the next two most closely related sequences are searched, until the whole tree has been constructed.

The clustalw program uses this more advanced method to create a guide-tree for multiple alignments.

7.3 Proteome clustering

We also used clustering is to produce protein families given only the proteome of an organism. In this application, the phylogenetic tree as a whole is not of primary interest, as it is too big to be of use. Therefore, an alternative approach to clustering is used. The first step is, as above, to calculate the alignment score matrix of all proteins against each others using the Needleman-Wunsch algorithm, described in Sections 3.1. By simply grouping all sequences above a suitable threshold, protein families can be created. The program checks each cell for similarity values above the threshold and joins the two sequences. If one or both of the sequences were already in a group, these are also joined. The global threshold has to be low enough to find all members of a protein family. This comes at the cost of joining protein families that should be distinct, but have some members containing similar domains. This problem is solved by statistical analysis of each group's distance matrix (created in the first step), removing scores that significantly deviate from the mean distance in that group, while being above the global threshold.

7.4 Principal components analysis

This is not a clustering method, but as clustering is often used on multidimensional data by specifying a distance measure, principal components analysis (PCA) is an alternative method for the analysis of such data. By finding the eigenvectors and their eigenvalues of a data set, it can often be reduced in dimensionality to the principle component, the eigenvector with the highest eigenvalue[63]. See Figure 7 for an illustration. An eigenvector is a vector yielding a scaled vector of itself when multiplied with a square matrix. There are a number of algorithms for finding eigenvectors of a given matrix, which has as many of them as it has dimensions. The square matrix used to find the eigenvectors is the covariance matrix of the data set.

We used this method to find the HMM transporter profile contributing most to the observed distance of organisms (Chapter 6). This can be used to find transporter families characteristic to certain groups of organisms, like for example parasites.

References

[1] M. A. Ferguson, M. G. Low, and G. A. Cross, "Glycosyl-sn-1,2-dimyristylphosphatidylinositol is covalently linked to Trypanosoma brucei variant surface glycoprotein," *J Biol Chem*, vol. 260, pp. 14547–14555, Nov 1985.

[2] R. Hooke, *Micrographia: or, Some physiological descriptions of minute bodies made by magnifying glasses.* London: J. Martyn and J. Allestry, first ed., 1665.

[3] L. M. Smith, J. Z. Sanders, R. J. Kaiser,

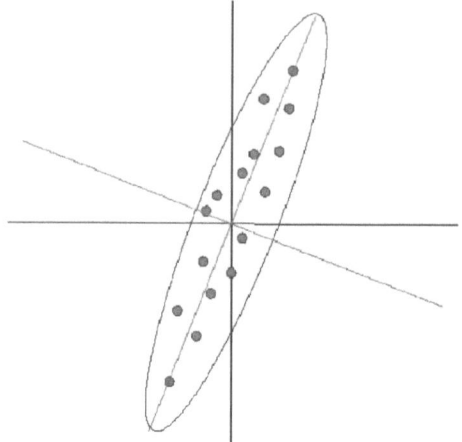

Figure 7: Points in 2-dimensional space[64]. The eigenvector with the highest eigenvalue (principle component) contributes most to the distribution of the points and goes from the lower left to the upper right quadrant.

P. Hughes, C. Dodd, C. R. Connell, C. Heiner, S. B. Kent, and L. E. Hood, "Fluorescence detection in automated DNA sequence analysis," *Nature*, vol. 321, pp. 674–679, Jun 1986.

[4] E. S. Lander, L. M. Linton, B. Birren, C. Nusbaum, M. C. Zody, J. Baldwin, *et al.*, "Initial sequencing and analysis of the human genome," *Nature*, vol. 409, pp. 860–921, Feb 2001.

[5] S. Anderson, "Shotgun DNA sequencing using cloned DNase I-generated fragments," *Nucleic Acids Res*, vol. 9, pp. 3015–3027, Jul 1981.

[6] M. Margulies, M. Egholm, W. E. Altman, S. Attiya, *et al.*, "Genome sequencing in microfabricated high-density picolitre reactors," *Nature*, vol. 437, pp. 376–380, Sep 2005.

[7] M. Ronaghi, S. Karamohamed, B. Pettersson, M. Uhlen, and P. Nyren, "Real-time DNA sequencing using detection of pyrophosphate release," *Anal Biochem*, vol. 242, pp. 84–89, Nov 1996.

[8] J. Shendure, G. J. Porreca, N. B. Reppas, X. Lin, J. P. McCutcheon, A. M. Rosenbaum, M. D. Wang, K. Zhang, R. D. Mitra, and G. M. Church, "Accurate multiplex polony sequencing of an evolved bacterial genome," *Science*, vol. 309, pp. 1728–1732, Sep 2005.

[9] J. R. Edwards, H. Ruparel, and J. Ju, "Mass-spectrometry DNA sequencing," *Mutat Res*, vol. 573, pp. 3–12, Jun 2005.

[10] J. Shine and L. Dalgarno, "Determinant of cistron specificity in bacterial ribosomes," *Nature*, vol. 254, pp. 34–38, Mar 1975.

[11] W. H. Majoros, M. Pertea, and S. L. Salzberg, "Tigrscan and glimmerhmm: two open source ab initio eukaryotic gene-finders," *Bioinformatics*, vol. 20, pp. 2878–2879, November 2004.

[12] C. Burge and S. Karlin, "Prediction of complete gene structures in human genomic dna.," *J Mol Biol*, vol. 268, pp. 78–94, April 1997.

[13] D. Kulp, D. Haussler, M. G. Reese, and F. H. Eeckman, "A generalized hidden Markov model for the recognition of human genes in DNA," *Proc Int Conf Intell Syst Mol Biol*, vol. 4, pp. 134–142, 1996.

[14] L. R. Rabiner, "A tutorial on hidden markov models and selected applications in speech recognition," pp. 267–296, 1990.

[15] S. Saxonov, P. Berg, and D. L. Brutlag, "A genome-wide analysis of CpG dinucleotides in the human genome distinguishes two distinct classes of promoters," *Proc Natl Acad Sci U S A*, vol. 103, pp. 1412–1417, Jan 2006.

[16] D. A. Benson, I. Karsch-Mizrachi, D. J. Lipman, J. Ostell, and D. L. Wheeler, "Genbank," *Nucleic Acids Research*, vol. 33, pp. D34+, January 2005.

[17] E. Gasteiger, A. Gattiker, C. Hoogland, I. Ivanyi, R. D. Appel, and A. Bairoch, "Expasy: The proteomics server for in-depth protein knowledge and analysis.," *Nucleic Acids Res*, vol. 31, pp. 3784–3788, July 2003.

[18] M. D. Adams, J. M. Kelley, J. D. Gocayne, M. Dubnick, M. H. Polymeropoulos, H. Xiao, C. R. Merril, A. Wu, B. Olde, and R. F. Moreno, "Complementary DNA sequencing: expressed sequence tags and human genome

project," *Science*, vol. 252, pp. 1651–1656, Jun 1991.

[19] C. D. Livingstone and G. J. Barton, "Protein sequence alignments: a strategy for the hierarchical analysis of residue conservation," *Comput Appl Biosci*, vol. 9, pp. 745–756, Dec 1993.

[20] S. Henikoff and J. G. Henikoff, "Amino acid substitution matrices from protein blocks," *Proc Natl Acad Sci U S A*, vol. 89, pp. 10915–10919, Nov 1992. Comparative Study.

[21] S. B. Needleman and C. D. Wunsch, "A general method applicable to the search for similarities in the amino acid sequence of two proteins," *J Mol Biol*, vol. 48, pp. 443–453, Mar 1970.

[22] T. F. Smith and M. S. Waterman, "Identification of common molecular subsequences," *J Mol Biol*, vol. 147, pp. 195–197, Mar 1981.

[23] S. F. Altschul, W. Gish, W. Miller, E. W. Myers, and D. J. Lipman, "Basic local alignment search tool," *J Mol Biol*, vol. 215, pp. 403–410, Oct 1990.

[24] J. D. Thompson, D. G. Higgins, and T. J. Gibson, "CLUSTAL W: improving the sensitivity of progressive multiple sequence alignment through sequence weighting, position-specific gap penalties and weight matrix choice," *Nucleic Acids Res*, vol. 22, pp. 4673–4680, Nov 1994. Comparative Study.

[25] N. Saitou and M. Nei, "The neighbor-joining method: a new method for reconstructing phylogenetic trees," *Mol Biol Evol*, vol. 4, pp. 406–425, Jul 1987.

[26] R. Chenna, H. Sugawara, T. Koike, R. Lopez, T. J. Gibson, D. G. Higgins, and J. D. Thompson, "Multiple sequence alignment with the Clustal series of programs," *Nucleic Acids Res*, vol. 31, pp. 3497–3500, Jul 2003.

[27] R. C. Edgar, "MUSCLE: multiple sequence alignment with high accuracy and high throughput," *Nucleic Acids Res*, vol. 32, no. 5, pp. 1792–1797, 2004. Comparative Study.

[28] R. C. Edgar, "MUSCLE: a multiple sequence alignment method with reduced time and space complexity," *BMC Bioinformatics*, vol. 5, p. 113, Aug 2004. Comparative Study.

[29] J. Heringa, "Detection of internal repeats: how common are they?," *Curr Opin Struct Biol*, vol. 8, pp. 338–345, Jun 1998.

[30] D. Worrall, L. Elias, D. Ashford, M. Smallwood, C. Sidebottom, P. Lillford, J. Telford, C. Holt, and D. Bowles, "A carrot leucine-rich-repeat protein that inhibits ice recrystallization," *Science*, vol. 282, pp. 115–117, Oct 1998.

[31] S. Labeit and B. Kolmerer, "Titins: giant proteins in charge of muscle ultrastructure and elasticity," *Science*, vol. 270, pp. 293–296, Oct 1995. Comparative Study.

[32] Y. Muragaki, N. Abe, Y. Ninomiya, B. R. Olsen, and A. Ooshima, "The human alpha 1(XV) collagen chain contains a large amino-terminal non-triple helical domain with a tandem repeat structure and homology to alpha 1(XVIII) collagen," *J Biol Chem*, vol. 269, pp. 4042–4046, Feb 1994. Comparative Study.

[33] L. Breeden and K. Nasmyth, "Similarity between cell-cycle genes of budding yeast and fission yeast and the Notch gene of Drosophila," *Nature*, vol. 329, pp. 651–654, Oct 1987.

[34] G. Parraga, S. J. Horvath, A. Eisen, W. E. Taylor, L. Hood, E. T. Young, and R. E. Klevit, "Zinc-dependent structure of a single-finger domain of yeast ADR1," *Science*, vol. 241, pp. 1489–1492, Sep 1988.

[35] J. Heringa and P. Argos, "A method to recognize distant repeats in protein sequences," *Proteins*, vol. 17, pp. 391–341, Dec 1993.

[36] M. S. Waterman and M. Vingron, "Rapid and accurate estimates of statistical significance for sequence data base searches," *Proc Natl Acad Sci U S A*, vol. 91, pp. 4625–4628, May 1994.

[37] M. Pellegrini, E. M. Marcotte, and T. O. Yeates, "A fast algorithm for genome-wide analysis of proteins with repeated sequences," *Proteins*, vol. 35, pp. 440–446, Jun 1999.

[38] A. Heger and L. Holm, "Rapid automatic detection and alignment of repeats in protein sequences," *Proteins*, vol. 41, pp. 224–237, Nov 2000.

[39] R. A. George and J. Heringa, "The repro server: finding protein internal sequence repeats through the web.," *Trends Biochem Sci*, vol. 25, pp. 515–517, October 2000.

[40] N. Fankhauser, T.-M. Nguyen-Ha1, J. Adler, and P. Maeser, "Surface antigens and potential virulence factors from parasites detected by comparative genomics of amino acid repeats," *Manuscript in revision*, 2007.

[41] D. P. Depledge, R. P. J. Lower, and D. F. Smith, "Repseq - a database of amino acid repeats present in lower eukaryotic pathogens," *BMC Bioinformatics*, vol. 8, pp. 122+, April 2007.

[42] A. D. McLachlan and J. Karn, "Periodic features in the amino acid sequence of nematode myosin rod," *J Mol Biol*, vol. 164, pp. 605–626, Mar 1983.

[43] M. V. Katti, R. Sami-Subbu, P. K. Ranjekar, and V. S. Gupta, "Amino acid repeat patterns in protein sequences: their diversity and structural-functional implications," *Protein Sci*, vol. 9, pp. 1203–1209, Jun 2000.

[44] R. Durbin, S. R. Eddy, A. Krogh, and G. Mitchison, *Biological Sequence Analysis : Probabilistic Models of Proteins and Nucleic Acids*. Cambridge University Press, July 1999.

[45] A. Markov, "Extension of the limit theorems of probability theory to a sum of variables connected in a chain," in *Dynamic Probabilistic Systems (Volume I: Markov Models)* (R. Howard, ed.), ch. Appendix B, pp. 552–577, New York City: John Wiley & Sons, Inc., 1971.

[46] A. Viterbi, "Error bounds for convolutional codes and an asymptotically optimum decoding algorithm," *Information Theory, IEEE Transactions on*, vol. 13, no. 2, pp. 260–269, 1967.

[47] L. E. Baum, T. Petrie, G. Soules, and N. Weiss, "A maximization technique occurring in the statistical analysis of probabilistic functions of markov chains," *The Annals of Mathematical Statistics*, vol. 41, no. 1, pp. 164–171, 1970.

[48] G. Florez-Larrahondo, S. Bridges, and E. A. Hansen, "Incremental estimation of discrete hidden markov models based on a new backward procedure," in *AAAI* (M. M. Veloso and S. Kambhampati, eds.), pp. 758–763, AAAI Press / The MIT Press, 2005.

[49] A. Krogh, M. Brown, I. S. Mian, K. Sjolander, and D. Haussler, "Hidden Markov models in computational biology. Applications to protein modeling," *J Mol Biol*, vol. 235, pp. 1501–1531, Feb 1994. Comparative Study.

[50] H. Abdi, "A neural network primer," 1994.

[51] V. M. Krasnopolsky, W. H. Gemmill, and L. C. Breaker, "A neural network multiparameter algorithm for ssm/i ocean retrievals - comparisons and validations," *Remote Sensing of Environment*, vol. 73, pp. 133–142, August.

[52] B. Rost, "PHD: predicting one-dimensional protein structure by profile-based neural networks," *Methods Enzymol*, vol. 266, pp. 525–539, 1996. Comparative Study.

[53] J. Meiler, M. Mueller, A. Zeidler, and F. Schmaeschke, "Generation and evaluation of dimension-reduced amino acid parameter representations by artificial neural networks," *J Mol Model*, no. 7, pp. 360–369, 2001.

[54] G. Bologna, C. Yvon, S. Duvaud, and A.-L. Veuthey, "N-Terminal myristoylation predictions by ensembles of neural networks," *Proteomics*, vol. 4, pp. 1626–1632, Jun 2004.

[55] T. Kohonen, *Self-Organizing Maps*, vol. 30 of *Springer Series in Information Sciences*. Berlin: Springer, 3 ed., 2001.

[56] T. Abe, S. Kanaya, M. Kinouchi, Y. Ichiba, T. Kozuki, and T. Ikemura, "Informatics for unveiling hidden genome signatures," *Genome Res*, vol. 13, pp. 693–702, Apr 2003.

[57] J. Aires-de Sousa and L. Aires-de Sousa, "Representation of DNA sequences with virtual potentials and their processing by (SEQREP) Kohonen self-organizing maps," *Bioinformatics*,

vol. 19, pp. 30–36, Jan 2003. Evaluation Studies.

[58] E. A. Ferran and P. Ferrara, "Clustering proteins into families using artificial neural networks," *Comput Appl Biosci*, vol. 8, pp. 39–44, Feb 1992.

[59] N. Fankhauser and P. Maser, "Identification of GPI anchor attachment signals by a Kohonen self-organizing map," *Bioinformatics*, vol. 21, pp. 1846–1852, May 2005.

[60] R. Sokal and C. Michener, "A statistical method for evaluating systematic relationships," *University of Kansas Science Bulletin*, vol. 38, pp. 1409–1438, 1958.

[61] M. B. Eisen, P. T. Spellman, P. O. Brown, and D. Botstein, "Cluster analysis and display of genome-wide expression patterns," *Proc Natl Acad Sci U S A*, vol. 95, pp. 14863–14868, Dec 1998.

[62] R. Plackett, "The discovery of the method of least squares," *Biometrika*, vol. 59, pp. 239–251, 1972.

[63] L. I. Smith, "A tutorial on principal components analysis," 2002.

[64] D. Schneegass, "Lektion zur principal component analysis und zur independent component analysis," 2004.

Chapter 3

Identification of GPI anchor attachment signals by a Kohonen self-organizing map

Sequence analysis

Identification of GPI anchor attachment signals by a Kohonen self-organizing map

Niklaus Fankhauser and Pascal Mäser*

Institute of Cell Biology, University of Bern, CH-3012 Bern, Switzerland

Received on September 29, 2004; revised on January 25, 2005; accepted on January 27, 2005
Advance Access publication February 2, 2005

ABSTRACT

Motivation: Anchoring of proteins to the extracytosolic leaflet of membranes via C-terminal attachment of glycosylphosphatidylinositol (GPI) is ubiquitous and essential in eukaryotes. The signal for GPI-anchoring is confined to the C-terminus of the target protein. In order to identify anchoring signals *in silico*, we have trained neural networks on known GPI-anchored proteins, systematically optimizing input parameters.

Results: A Kohonen self-organizing map, GPI-SOM, was developed that predicts GPI-anchored proteins with high accuracy. In combination with SignalP, GPI-SOM was used in genome-wide surveys for GPI-anchored proteins in diverse eukaryotes. Apart from specialized parasites, a general trend towards higher percentages of GPI-anchored proteins in larger proteomes was observed.

Availability: GPI-SOM is accessible on-line at http://gpi.unibe.ch. The source code (written in C) is available on the same website.

Contact: pascal.maeser@izb.unibe.ch

Supplementary information: Positive training set, performance test sets and lists of predicted GPI-anchored proteins from different eukaryotes in fasta format.

INTRODUCTION

Anchoring of proteins to the extracellular surface of the plasma membrane via glycosylphosphatidylinositol (GPI) is widespread among eukaryotes. GPI-anchored proteins range from small peptides to large antigens and fulfill a variety of cellular functions. Some are receptors for external signals, e.g. Nogo receptor (Fournier *et al.*, 2001) or Trail decoy receptors (Sheridan *et al.*, 1997), others for nutrients such as the folate receptor (Lacey *et al.*, 1989). Extracellular proteases and other enzymes may be GPI-anchored (Netzel-Arnett *et al.*, 2003). Structural surface proteins with a GPI anchor are of particular importance as antigens of eukaryotic parasites (Ferguson, 1999). There are also GPI-anchored proteins of unknown function, such as the prion protein involved in bovine spongiform encephalopathy (Stahl *et al.*, 1987). GPI-anchoring is essential for cell function and development, indicated by the fact that null mutations in GPI synthesis are lethal to the yeast *Saccharomyces cerevisiae* (Hamburger *et al.*, 1995; Sutterlin *et al.*, 1998). Mice lacking GPI synthesis fail in their development at early embryonic stages (Nozaki *et al.*, 1999).

Proteins destined to receive a GPI-anchor carry a C-terminal signal sequence. This sequence is sufficient for GPI-anchor attachment,

*To whom correspondence should be addressed.

as has been demonstrated by gene fusion experiments (Caras *et al.*, 1987). Furthermore, heterologous expression systems revealed that the GPI-anchor attachment signal is generally recognized across eukaryotic kingdoms, though not necessarily in all instances (Moran and Caras, 1994; Meyer *et al.*, 2002). Signal sequences were functional from *Pneumocystis carinii* in COS cells (Guadiz *et al.*, 1998), from *Homo sapiens* in *Trypanosoma brucei* (Butikofer *et al.*, 1999) and from rat in *Pichia pastoris* (Morel and Massoulie, 1997). However, the C-termini from known GPI-anchored proteins cannot be aligned to a consensus sequence. The GPI anchor attachment signal is cleaved during protein processing and the preassembled GPI core structure is covalently attached to the new C-terminus of the target protein, termed omega (ω) site (Takeda and Kinoshita, 1995). Since these reactions take place in the lumen of the endoplasmic reticulum (ER), a C-terminal GPI anchor-attachment signal only makes sense in the context of an N-terminal export sequence. The canonical tool for prediction of the latter type of signal is SignalP, a program that uses hidden Markov models and a neural network (Nielsen *et al.*, 1997). Two programs are available for computational prediction of C-terminal GPI-anchoring signals, Big-PI (Eisenhaber *et al.*, 1999, http://mendel.imp.univie.ac.at/sat/gpi/gpi_server.html) and DGPI (Kronegg and Buloz, 1999, http://129.194.185.165/dgpi/). Both are based on the amino acid composition around the ω site (Udenfriend and Kodukula, 1995; Eisenhaber *et al.*, 1998). Such programs are most useful when predicting the ω site of proteins known to be GPI-anchored. For screening of unknown proteins, however, it is difficult to balance between false positive and false negative errors. Big-PI now exists in kingdom-specific flavors (http://mendel.imp.univie.ac.at/gpi/gpi_server.html for animals or protozoa, http://mendel.imp.univie.ac.at/gpi/fungi_server.html for fungi, http://mendel.imp.univie.ac.at/gpi/plant_server.html for plants).

Neural networks of the Kohonen type, also termed self-organizing maps (SOMs), are powerful tools for classification of hidden information in large datasets (Kohonen, 2001). As with classical feed-forward networks, learning in SOMs happens by adjusting the weights of the connections (synapses) between units (neurons). But in contrast to feed-forward nets, SOMs learn in an unsupervised manner, guaranteeing minimal bias from the investigator. Thus SOMs will distinguish patterns without knowing if and how many different patterns the input contains. Furthermore, SOM output can easily be visualized as a two-dimensional map. Biological applications range from clustering of microarray data (Toronen *et al.*, 1999) to analysis of whale songs (Murray *et al.*, 1998). SOMs have successfully been applied for classification of DNA sequences based on codon usage (Kanaya et al., 2001) (Supek and Vlahovicek, 2004), nucleotide frequencies (Abe et al., 2003), or virtual potentials (Aires-de-Sousa and Aires-de-Sousa, 2003). SOM analysis of protein sequences was carried out using bipeptide composition as input (Ferran and Ferrara, 1992; Ferran et al., 1994).

Encouraged by the facts that the GPI anchor attachment signal (1) carries universal features and (2) is confined to the C-terminus of the target protein, we implemented neural network approaches for identification of GPI-anchoring signals. Here, we present a case study for development and systematic optimization of a SOM that recognizes GPI-anchored proteins from diverse eukaryotes.

SYSTEMS AND METHODS

Hardware

The University of Bern Linux cluster Ubelix (http://ubelix.unibe.ch) was used for running multiple experiments in parallel in order to optimize network architecture and input parameters. The final program GPI-SOM and its web interface (http://gpi.unibe.ch) are running on an AMD64 gentoo Linux server.

Neural networks

All neural networks were implemented with the artificial neural network library (ANNLIB) (A.Hoekstra, M.A.Kraaijveld, D.de Ridder, W.F.Schmidt, Pattern Recognition Group, Delft University of Technology) and written in C. PNG image files of two-dimensional maps were generated using the GD graphics library (http://www.Boutell.com). The web interface was written in Perl-cgi.

Training and evaluation sets

The positive training and evaluation sets consisted of proteins that had been experimentally shown to be GPI-anchored. These included 110 proteins of all four eukaryote kingdoms selected via Entrez from GenBank, supplemented with a set of 248 GPI-proteins from *Arabidopsis thaliana*, kindly provided by P.Dupree, University of Cambridge (Borner et al., 2003). The positive test sets for Table 2 were (e) a list of GPI-anchored proteins downloaded from the website of B.Eisenhaber, University of Vienna (http://mendel.imp.univie.ac.at/gpi/gpi.p/gpi.swp), excluding those already present in our positive training and validation sets, and (f) recently published, experimentally verified GPI-anchored proteins that none of the tested programs had encountered before.

The negative training and evaluation sets consisted of 256 known cytosolic and 128 transmembrane proteins of all eukaryote kingdoms, 25 of which had a transmembrane domain near their C-terminus. The negative test sets for Table 2 were selected from GenBank by text-based searches. For the set N–TM–C, only transmembrane proteins with an N-terminal export signal predicted by SignalP as well as a hydrophobic C-terminus were selected. All protein sets were homology-reduced with a Perl script that uses the Smith/Waterman algorithm (Smith and Waterman, 1981) to find any two sequences that have an alignment score above a certain percentage of the shorter sequence's selfmatch score. The shorter sequence will be removed in order to create a set of non-homologous proteins. The threshold for sequence removal was set to 50% for the negative set and 80% for the C-terminal 32 amino acids of GPI anchored proteins.

Random sequences between 80 and 400 amino acids in length (random distribution) were generated based on the amino acid frequencies of the predicted *S.cerevisiae* proteome (A, 0.055; C, 0.013; D, 0.058; E, 0.065; F, 0.045; G, 0.050; H, 0.022; I, 0.066; K, 0.073; L, 0.096; M, 0.021; N, 0.061; P, 0.043; Q, 0.039; R, 0.045; S, 0.090; T, 0.059; V, 0.056; W, 0.010; Y, 0.034), with a Perl script utilizing random numbers from http://random.org.

Table 1. Selected formats of sequence representation, their corresponding numbers of input residues (AAs), numbers of cells in the input layer and their performance as indicated by validation error (FP, false positives; FN, false negatives) of feed forward networks trained by back-propagation

Interface	AAs	Input cells	FN (%)	FP (%)
2D	32	640	3.1	3.2
H	32	32	4.7	12
VP	32	20	13	15
VP + H	32	52	1.6	7.2
Z	32	20	3.1	6.4
$Z + H$	32	52	3.9	3.2
$Z + H + \omega$	32	54	3.1	2.4
$Z + H + \omega$	22	44	3.1	1.6

Input elements: 2D, two-dimensional interface; H, hydrophobicity; VP, virtual potential; Z, zentriole; ω, omega site.

Proteome files

Predicted proteins from completely sequenced eukaryotic genomes were obtained from ftp.ebi.ac.uk (*A.thaliana, Drosophila melanogaster, S.cerevisiae, Schizosaccharomyces pombe*), ftp.ensembl.org (*Caenorhabditis elegans, H.sapiens, Anopheles gambiae*), ftp.ncbi.nlm.nih.gov (*Encephalitozoon cuniculi, Mus musculus*), ftp.sanger.ac.uk (*T.brucei* chromosome 1), ftp.tigr.org (*T.brucei* chromosome 2), and www.plasmodb.org (*Plasmodium falciparum*).

ALGORITHMS

Network architecture and training

Pilot experiments for optimizing input parameters were run as feed-forward networks for sake of speed. These networks contained variable numbers of input units (depending on input format; Table 1), one hidden layer of 10 units, a second one of 5, and 2 units in the output layer. These networks were trained by back-propagation with a constant learning rate of 0.001 (a gradually decreasing learning rate was tried out but did not perform better). The weights of all connections were initially set at random. After each round of training, all weights were updated by back-propagation and saved to a separate file. After 5000 rounds, weight values yielding minimal validation error were restored to avoid over-training of the network (i.e. minimizing training error at the cost of validation error; Kohonen, 2001). Protein sets had been split 2:1 training to validation.

Kohonen SOMs were also trained for 5000 rounds starting from random weights, but updating of weights was restricted to the winning unit and its neighbors (radius scaled by the Gaussian function of distance). After each cycle, the winning units were determined for the validation sets and the number of units responding to sequences from both positive and negative sets was taken as a negative measure of quality. The map was saved only when the number of such undecided units was lower than in any previous step. Thus, upon completion of training, the network had been stored optimized with respect to validation. For visual evaluation, every unit was represented as a colored square according to class and intensity representing how often a particular unit had won.

Sequence representation

A number of different input formats were investigated (see Implementation section). Virtual potentials (VP) for amino acids were calculated in analogy to the formula proposed for DNA sequences (Aires-de-Sousa and Aires-de-Sousa, 2003). The VP at the C-terminal position of a preceding window sized 32 was used as input. For three occurrences of amino acid A at positions p_{A1}, p_{A2}, p_{A3}, the VP equals $((p_{A1})^{-1} + (p_{A2})^{-1} + (p_{A3})^{-1})$, where p counts upwards from 1 starting at distance 32 from the C-terminus. The zentriole Z of a given amino acid A represents its average position weighed by its proximity to the C-terminus. For three occurrences of A at positions p_{A1}, p_{A2}, p_{A3} counted upwards from 1 starting at distance 32 from the C-terminus, Z was defined as $((p_{A1}/2 + p_{A2})/2 + p_{A3})/2$, which generalizes to

$$Z(A) = 2^{-n} \sum_{i=1}^{n} 2^{i-1} p_{Ai}. \quad (1)$$

For amino acids not occurring in the input sequence, Z equals zero. The quality of a putative omega site was assessed by a scoring matrix for the triplet $\omega, \omega + 1, \omega + 2$, based on known ω sites (Gerber et al., 1992; Kodukula et al., 1993; Udenfriend and Kodukula, 1995; Eisenhaber et al., 1998). Top scores were attributed to serine followed by alanine and glycine. Hydrophobicity scores of amino acids were derived from Kyte and Doolittle (1982).

Automated filling up of the map

Empty units in a SOM that had not been hit during training were classified according to their surroundings. Scores for GPI and non-GPI of all units within a radius of three around the empty one were multiplied with a distance factor (3, 1.5, or 1 beginning with the innermost layer) and summed up. If the difference between the two sums was >1, the unit was assigned to the higher-scoring class; otherwise it was left undecided.

IMPLEMENTATION

Optimizing sequence representation

Transformation of biological sequence data into a form that can be read by the input layer of a neural network inevitably causes a substantial loss of information, since it is not practicable to express molecular structure in numbers. We have evaluated different numerical representation formats of amino acid sequences for identification of GPI proteins from their 32 C-terminal residues. Beginning with collinear versions, where input neurons directly represent individual amino acid positions, a two-dimensional interface of 20 binary input units for each of the 32 positions was tried. The resulting network performed with an accuracy of ~97%, but it was impractical because of the large amounts of data and long computation times (Table 1). Computation was accelerated by representing each position with a single unit instead of twenty; in that case, amino acids were substituted by their relative hydrophobicity (Kyte and Doolittle, 1982). However, this increased the number of wrong predictions, particularly false positive ones (Table 1). Addition of an input unit for the local alignment score to a reference GPI signal sequence (the last 31 amino acids of pig renal dipeptidase, GenBank accession P22412) did not reduce validation errors (not shown).

Virtual potentials have been used for positional transformation of DNA sequences (Aires-de-Sousa and Aires-de-Sousa, 2003). We have adapted this concept to amino acids. This transformation obviously reduced input size and computation time compared to collinear representations, but resulted in only ~85% correct predictions.

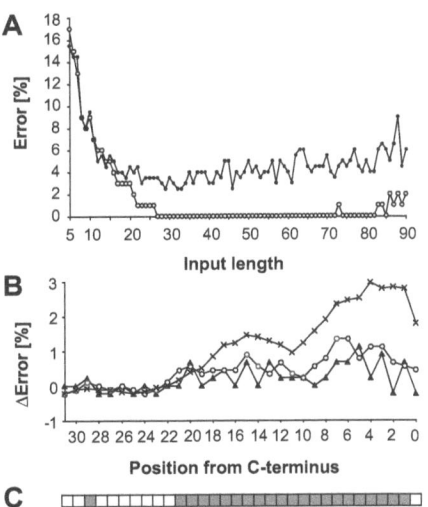

Fig. 1. Selection of input residues from the C-terminus with feed-forward networks. (A) Prediction accuracy in function of input length. The average percentage of false positives and false negative predictions on the training sets (white circles) and validation sets (black circles) is plotted against the number of amino acids counted from the C-terminus. Validation error was minimal at an input size of 29. (B) Simulated mutagenesis of the presumed signal. Single positions (black triangles), pairs (white circles), or groups of four amino acids (crosses) were masked sequentially and the performance of the network was evaluated as average of positive and negative validation errors between masked and original input sequence. (C) 22 important positions (filled squares) were used as input for the Kohonen map GPI-SOM.

These high error rates were, however, substantially reduced by the addition of input units for relative hydrophobicity (H) at each position (Table 1). Thus the combination of a positionally transformed parameter (VP) for each of the 20 amino acids with a collinear representation (H) for each position of the C-terminus appeared to be a suitable input format for recognizing GPI-anchored proteins, while neither VP nor H alone performed well. Related to the virtual potential is the concept of the zentriole (Z), a C-proximally weighed average position (described under Algorithms). Already by itself, the zentriole input format performed promisingly well and combined with hydrophobicity values of each position, it achieved minimal error rates. Further studies and optimization were, therefore, carried out with this type of input vector ($Z + H$).

Narrowing down on the signal sequence

In order to streamline input data in respect to signal recognition, a fast feed-forward network was repeatedly trained and evaluated with C-terminal fragments from the GPI positive sets, each time increasing the length of input sequences by one (Fig. 1A). Initially, both training error and validation error decreased with increasing length of input sequence, reaching a minimum at 29 amino acids.

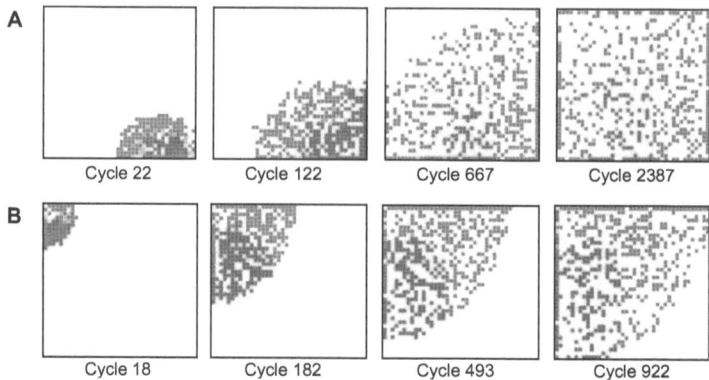

Fig. 2. Unsupervised learning by self-organizing Kohonen maps. Using the input residues outlined in Figure 1C, Kohonen SOMs of size 40 × 40 were trained with different amino acid representation formats. (**A**) Representing each residue with its Kyte-Doolittle hydrophobicity resulted in poor separation and slow convergence of self-organization. (**B**) Hydrophobicity of each residue combined with the zentriole Z for each amino acid performed much better. Clear and fast separation of GPI-proteins and non-GPI proteins was observed. Note that coloring took place after training; during self-organization, the SOM is not told which sequences are GPI and which are not. Color intensity indicates how often a particular unit was activated (green, GPI; red, non-GPI; yellow, activated by members of either set).

From 32 residues upwards, however, the validation error rose again, indicative of excessive information. Therefore only the 32 C-terminal amino acids were selected as input for further analyses.

By performing an *in silico* mutagenesis experiment, we investigated which of the 32 C-terminal residues best distinguished a GPI-anchored protein as such. A sliding window that represented any amino acid as 'X' was used to mask each position in turn (X was assigned the hydrophobicity of alanine). As expected, prediction accuracy decreased with increasing window size (Fig. 1B). The amino acids far from the C-terminus were, with a few exceptions, less significant than the ones near it (Fig. 1B). Based on these data, positions to be presented to the network were selected and the most efficient combination was identified empirically. It was an input vector of 22 residues (Fig. 1C) which, when fed into the network, performed even better than the vector of all 32 C-terminal amino acids (Table 1).

The most frequent source of false positives were integral membrane proteins with a transmembrane domain within the last 30 amino acids. In order to better distinguish GPI-anchoring signals from transmembrane domains, two extra units were added to the input layer: one for the quality of a putative ω site and one for its position. This further decreased error rates (Table 1). Thus, the final input vector contained 44 components ($Z + H + \omega$; Table 1).

GPI-SOM

The final GPI anchoring signal prediction program GPI-SOM was implemented as a Kohonen SOM with an input layer of 44 neurons as described above. Square output maps of side length 10 did not provide enough room for both classes to segregate (not shown). With increasing side length there was a clearer separation of GPI and non-GPI proteins, until at length 40 the number of units in the map that were excited by proteins from both positive and negative sets was minimal. Figure 2 shows the process of self-organization during training. After a few cycles it became evident that the classes were separating using the zentriole plus hydrophobicity input vector ($Z + H$; Fig. 2B), illustrating that prediction of GPI-anchoring is solvable with a SOM. The collinear hydrophobicity vector alone did not distinguish clearly between GPI-positive and -negative proteins and the SOM took longer to reach minimal ambiguity (H; Fig. 2A).

After training, blank units in the map were classified based on their surroundings (see Algorithms). Since there were more than twice as many units in the SOM than sequences in the training sets, the majority of units was assigned only after training. While the units inside the GPI (blue) and non-GPI (green) regions were straightforward to assign, 11 of the units in between the two areas had to be left 'undecided' (red in Fig. 3). If such a blank unit is hit by a test sequence, there will be no prediction made (classified 'uncertain'). Furthermore, there was an inactive region of 185 blank units at the edge of the map that no input sequence has activated so far (Fig. 3). GPI-SOM is accessible via http://gpi.unibe.ch and accepts batch input in fasta format.

Evaluation of different GPI-prediction programs

A series of positive and negative test sets consisting of proteins from all eukaryote kingdoms were used to assess sensitivity and selectivity of the GPI-anchoring prediction programs BigPI, DGPI, GPI-SOM, and its corresponding feed-forward network ($Z + H + \omega$). Since a target protein must have an N-terminal ER export signal to receive its GPI anchor all programs were combined with SignalP (HMM version; Nielsen and Krogh, 1998), except for DGPI which already considers the N-terminus of the target protein. Prediction of GPI-anchored proteins based on their C-termini alone is not sensible since GPI-anchoring signals are only meaningful inside the ER (a presumed C-terminal GPI anchor attachment sequence, even a perfect one, is meaningless in the absence of an N-terminal export sequence).

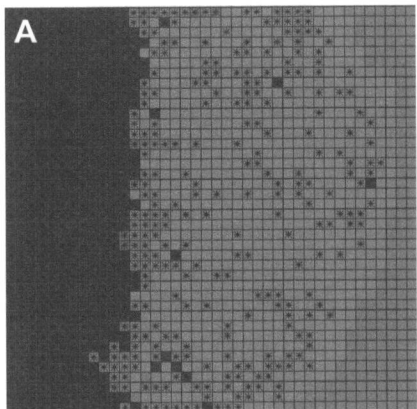

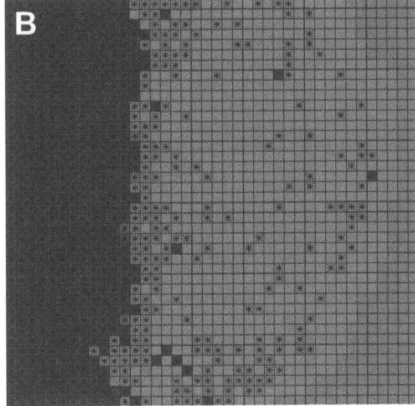

Fig. 3. The final map GPI-SOM. The map of 40 × 40 units was filled completely as described in Algorithms, and subdivided into three types of fields: GPI (green), non-GPI (blue) and undecided (red). This allowed fast scanning of large datasets. Black dots represent hits for (**A**) the predicted proteome of S.cerevisiae (5864 proteins) and (**B**) the same number of random sequences of the same amino acid frequencies as S.cerevisiae proteins. Intensity indicates how often a unit was hit. In the online version (http://gpi.unibe.ch) each unit is clickable, producing a list of the proteins that activated it.

Table 2. Performance of GPI-anchoring prediction programs

	BigPI	DGPI	FF	GPI-SOM
Negative sets				
(a) Cytosolic	0	1.5	2.0	0.5
(b) Secreted	0	1.5	2.9	1.5
(c) TM	0	0	2.5	0.6
(d) N–TM–C	1.9	27	32	34
Positive sets				
(e) GPI	17	17	4	4
(f) new GPI	48	14	2.4	4.8

GPI-SOM and its corresponding feed-forward network (FF) are compared to Big-PI (Eisenhaber et al., 2003) and DGPI (Kronegg and Buloz, 1999) using different test sets: (a) cytosolic proteins (196 sequences); (b) secreted proteins (68 sequences); (c) transmembrane proteins (159 sequences); (d) transmembrane proteins with N-terminal export signal and hydrophobic C-terminus (107 sequences); (e) GPI-anchored proteins not present in our positive training and validation sets (75 sequences); (f) recently published GPI-proteins which none of the programs had seen before (42 sequences). All test sets are available as supplementary material. BigPI, FF and GPI-SOM were combined with the HMM output of SignalP; DGPI already considers the N-terminus on its own. Numbers are the percentage of false predictions.

Thus only proteins predicted to have both N- and C-terminal signals were classified as GPI-anchored.

As shown in Table 2, Big-PI was extremely specific, with hardly any false positive predictions throughout the negative test sets. The other programs also performed well, except against transmembrane proteins with an N-terminal export sequence plus a hydrophobic C-terminus (row d). These are the proteins most closely resembling GPI-anchored ones (Dalley and Bulleid, 2003) and, accordingly, the false positive error rates were around 30%. Regarding sensitivity, the feed forward network and GPI-SOM performed best. BigPI exhibited the highest rate of false negative predictions, presumably the price for its excellent specificity.

Genome-wide surveys for GPI-anchored proteins

GPI-SOM combined with SignalP was used in genome-wide surveys for GPI-anchored proteins in a number of eukaryotes. The S.cerevisiae proteome is shown as an example in Figure 3A. Of the total 5864 sequences, GPI-SOM predicted 438 positives, 121 of which were assigned N-terminal signals by SignalP resulting in 2.1% predicted GPI-anchored proteins. As stated above, the 307 proteins with predicted C-terminal signal but lacking an N-terminal one cannot be classified false positives; such predictions are meaningless (in order to test C-terminal predictions experimentally, the respective proteins would need to be fused to an ER export signal). For comparison, 5864 random sequences of the same amino acid frequencies as yeast proteins are shown in Figure 3B. GPI-SOM predicted 437 positives, of which only 8 (0.14%) were also predicted to possess an N-terminal export sequence by SignalP.

Most organisms appeared to have between 2 and 3% GPI-anchored proteins. Notable exceptions were *E.cuniculi* with only 0.5% and *T.brucei* with 5.6% predicted GPI-proteins (Fig. 4). Both are highly specialized parasites. Among the remaining organisms, a trend was observed toward a higher percentage of GPI-anchored proteins in organisms with larger proteomes (Fig. 4). Lists of predicted GPI-anchored proteins for different organisms are available as Supplementary information or from the GPI-SOM website.

DISCUSSION

SOMs are powerful tools for the detection and the classification of hidden patterns, but applications to proteins are hampered by the size of input data and by the inherent problem that conversion of

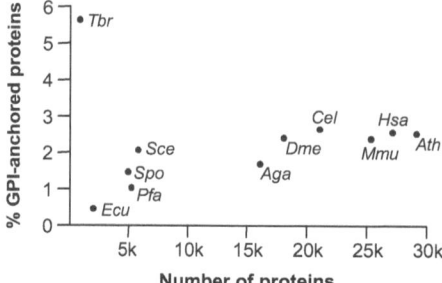

Fig. 4. Genome-wide predictions of GPI proteins. Predicted proteomes of completely sequenced eukaryotic genomes were screened for presumably GPI-anchored proteins with GPI-SOM. Since anchoring signals are only meaningful in the ER, only hits for which SignalP predicted an N-terminal export signal were counted as GPI-proteins. Organisms with more genes tended to have a higher percentage of GPI-anchored proteins (Aga, *A.gambiae*; Ath, *A.thaliana*; Cel, *C.elegans*; Dme, *D.melanogaster*; Ecu, *E.cuniculi*; Hsa, *H.sapiens*; Mmu, *M.musculus*; Pfa, *P.falciparum*; Sce, *S.cerevisiae*; Spo, *S.pombe*; Tbr; *T.brucei* chromosomes 1 and 2).

amino acid sequences to numerical format causes substantial loss of chemical information. Signal sequences localized within proteins, however, may be suitable targets for neural networks (Nielsen *et al.*, 1997). This has been demonstrated by the good performance of the feed-forward network SignalP in predicting N-terminal export sequences (Bendtsen *et al.*, 2004). Here we present GPI-SOM, a self-organizing map that recognizes C-terminal GPI-anchor attachment signals with good accuracy. It was developed by systematic, target-oriented optimization of input parameters and network architecture. Input consists of 44 numbers: the zentrioles for each amino acid (20 units), hydrophobicity of selected C-terminal positions (22 units), and quality and position of the best match for a putative ω site (2 units). The zentriole represents the average position of a given amino acid weighed by C-terminal proximity, a transformation somewhat related to the concept of virtual potentials (Aires-de-Sousa and Aires-de-Sousa, 2003). The output layer is a square map of 1600 units, where anchored and non-anchored proteins clearly separate (Fig. 2). The map was finalized by an algorithm that categorized empty or ambiguous units based on their surroundings. The good performance of GPI-SOM indicates that, in principle, the problem of GPI-anchoring signal prediction is solvable with a SOM.

GPI-SOM had a sensitivity of ~0.96 (Table 2). Selectivity is less straightforward to assess and depends greatly on the nature of the negative test proteins (Table 2). The main source of false positive predictions were integral membrane proteins with a transmembrane domain at their C-terminus (Table 2, row d). This is an inherent problem with GPI-anchoring signals; indeed, it has been shown experimentally that one point mutation may suffice to convert an anchor attachment signal to a transmembrane domain (Dalley and Bulleid, 2003). Misinterpretation of integral membrane proteins for anchored ones might be minimized by excluding sequences with multiple predicted transmembrane domains. However, we refrained from doing so since it cannot be excluded on the assumption that a protein has a C-terminal GPI anchor in addition to internal transmembrane domains.

A drawback of neural networks is that the machine is learning but not the investigator. In most cases, it is impossible to track the connections of a trained network and determine which input features are the most important. We have circumvented this problem by systematically altering input data. Thus, varying the length of input sequence (Fig. 1A) followed by a simulated mutagenesis experiment (Fig. 1B) identified crucial positions in the C-terminus distinguishing GPI-anchored from non-anchored proteins. This allowed maximal prediction accuracy with minimal input data (Table 1).

Surveys for GPI-anchored proteins were carried out in eukaryotes for which unbiased protein sets from completely sequenced chromosomes were available. Most species had between 2 and 3% predicted GPI-proteins (Fig. 4). Genome-wide prediction of GPI-proteins is critical because the error rates of GPI-SOM are in the same order of magnitude as the percentages of GPI-anchored proteins in a given proteome. Thus the predicted numbers of GPI-proteins have to be taken with caution. Also prediction of N-terminal signal sequences with SignalP, which is a prerequisite for prediction of GPI-anchor attachment sites, involves a certain error. Nevertheless, genome-wide comparisons between different species may yield insights into their use of GPI anchors. There appeared to be a trend towards higher percentages of GPI-anchored proteins in larger proteomes. No such trend was observed in transmembrane proteins (Ward, 2001). Top and bottom positions in Figure 4 were taken by the parasitic protozoa *T.brucei* (5.6% GPI-proteins) and *E.cuniculi* (0.5% GPI-proteins), respectively. *T.brucei* proliferate extracellularly in the mammalian bloodstream. Evading the host's immune system by variation of their surface coat, *T.brucei* spp. have a repertoire of several hundred genes for GPI-anchored surface glycoproteins (Donelson, 2003). The microsporidian *E.cuniculi*, in contrast, is an obligate intracellular parasite and might, therefore, not be expected to possess GPI-anchored proteins at all. However, GPI-SOM in combination with SignalP identified 9 candidate proteins with N- and C-terminal signals, among which a proteinase and proteins similar to oligosaccharide deacetylase, glucosyltransferase, and glucan glucosidase (see Supplementary data). *E.cuniculi* lacks several of the enzymes involved in GPI synthesis and attachment; but surprisingly, it has a predicted protein with high similarity to phosphatidylinositol N-acetylglucosaminyltransferase (GPI2), catalyzing the first step in GPI synthesis (GenBank accession NP_597633 has a p-value of 2e-119 against PFAM entry PF06432). Whether the nine *E.cuniculi* proteins predicted to receive an anchor are false positives or whether some of these proteins actually get anchored to the host cell membrane remains to be investigated.

In summary, GPI-SOM is a new approach towards computational prediction of GPI-anchoring signals and provides a welcome addition to the existing programs Big-PI and DGPI which predict GPI anchor attachment sites based on statistical expectation.

ACKNOWLEDGEMENTS

We wish to thank Isabel Roditi, Peter Bütikofer, Walter Senn and Daniel Stalder for the helpful advice, and Paul Dupree for the provision of sequences of GPI-anchored *Arabidopsis* proteins. This work was supported by a Swiss National Science Foundation research professorship grant to P.M.

REFERENCES

Abe,T. *et al.* (2003) Informatics for unveiling hidden genome signatures. *Genome Res.*, **13**, 693–702.

Aires-de-Sousa,J. and Aires-de-Sousa,L. (2003) Representation of DNA sequences with virtual potentials and their processing by (seqrep) Kohonen self-organizing maps. *Bioinformatics*, **19**, 30–36.

Bendtsen,J.D. *et al.* (2004) Improved prediction of signal peptides: SignalP 3.0. *J. Mol. Biol.*, **340**, 783–795.

Borner,G.H. *et al.* (2003) Identification of glycosylphosphatidylinositol-anchored proteins in *Arabidopsis*. A proteomic and genomic analysis. *Plant Physiol.*, **132**, 568–577.

Butikofer,P. *et al.* (1999) Phosphorylation of a major GPI-anchored surface protein of *Trypanosoma brucei* during transport to the plasma membrane. *J. Cell Sci.*, **112** (Pt 11), 1785–1795.

Caras,I.W. (1987) Signal for attachment of a phospholipid membrane anchor in decay accelerating factor. *Science*, **238**, 1280–1283.

Dalley,J.A. and Bulleid,N.J. (2003) The endoplasmic reticulum (ER) translocon can differentiate between hydrophobic sequences allowing signals for glycosylphosphatidylinositol anchor addition to be fully translocated into the ER lumen. *J. Biol. Chem.*, **278**, 51749–51757.

Donelson,J.E. (2003) Antigenic variation and the African trypanosome genome. *Acta Trop.*, **85**, 391–404.

Eisenhaber,B. *et al.* (1998) Sequence properties of GPI-anchored proteins near the omega-site: constraints for the polypeptide binding site of the putative transamidase. *Protein Eng.*, **11**, 1155–1161.

Eisenhaber,B. *et al.* (1999) Prediction of potential GPI-modification sites in proprotein sequences. *J. Mol. Biol.*, **292**, 741–758.

Eisenhaber,F. *et al.* (2003) Prediction of lipid posttranslational modifications and localization signals from protein sequences: Big-pi, nmt and pts1. *Nucleic Acids Res.*, **31**, 3631–3634.

Ferguson,M.A. (1999) The structure, biosynthesis and functions of glycosylphosphatidylinositol anchors, and the contributions of trypanosome research. *J. Cell Sci.*, **112** (Pt 17), 2799–2809.

Ferran,E.A. and Ferrara,P. (1992) Clustering proteins into families using artificial neural networks. *Comput. Appl. Biosci.*, **8**, 39–44.

Ferran,E.A. *et al.* (1994) Self-organized neural maps of human protein sequences. *Protein Sci.*, **3**, 507–521.

Fournier,A.E. *et al.* (2001) Identification of a receptor mediating nogo-66 inhibition of axonal regeneration. *Nature*, **409**, 341–346.

Gerber,L.D. *et al.* (1992) Phosphatidylinositol glycan (pi-g) anchored membrane proteins. Amino acid requirements adjacent to the site of cleavage and pi-g attachment in the COOH-terminal signal peptide. *J. Biol. Chem.*, **267**, 12168–12173.

Guadiz,G. *et al.* (1998) The carboxyl terminus of *Pneumocystis carinii* glycoprotein a encodes a functional glycosylphosphatidylinositol signal sequence. *J. Biol. Chem.*, **273**, 26202–26209.

Hamburger,D. *et al.* (1995) Yeast gaa1p is required for attachment of a completed GPI anchor onto proteins. *J. Cell Biol.*, **129**, 629–639.

Kanaya,S. *et al.* (2001) Analysis of codon usage diversity of bacterial genes with a self-organizing map (SOM): characterization of horizontally transferred genes with emphasis on the *E.coli* O157 genome. *Gene*, **276**, 89–99.

Kodukula,K. *et al.* (1993) Biosynthesis of glycosylphosphatidylinositol (GPI)-anchored membrane proteins in intact cells: specific amino acid requirements adjacent to the site of cleavage and GPI attachment. *J. Cell Biol.*, **120**, 657–664.

Kohonen,T. (2001) *Self-Organizing Maps*, 3rd edn, Springer Series in Information Sciences, Vol. 30, Springer, Berlin.

Kronegg,J. and Buloz,D. (1999) Detection/prediction of GPI cleavage site (GPI-anchor) in a protein (DGPI).

Kyte,J. and Doolittle,R.F. (1982) A simple method for displaying the hydropathic character of a protein. *J. Mol. Biol.*, **157**, 105–132.

Lacey,S.W. *et al.* (1989) Complementary DNA for the folate binding protein correctly predicts anchoring to the membrane by glycosyl-phosphatidylinositol. *J. Clin. Invest.*, **84**, 715–720.

Meyer,U. *et al.* (2002) The glycosylphosphatidylinositol (GPI) signal sequence of human placental alkaline phosphatase is not recognized by human gpi8p in the context of the yeast GPI anchoring machinery. *Mol. Microbiol.*, **46**, 745–748.

Moran,P. and Caras,I.W. (1994) Requirements for glycosylphosphatidylinositol attachment are similar but not identical in mammalian cells and parasitic protozoa. *J. Cell Biol.*, **125**, 333–343.

Morel,N. and Massoulie,J. (1997) Expression and processing of vertebrate acetylcholinesterase in the yeast *Pichia pastoris*. *Biochem. J.*, **328** (Pt 1), 121–129.

Murray,S.O. *et al.* (1998) The neural network classification of false killer whale (*Pseudorca crassidens*) vocalizations. *J. Acoust. Soc. Am.*, **104**, 3626–3633.

Netzel-Arnett,S. *et al.* (2003) Membrane anchored serine proteases: a rapidly expanding group of cell surface proteolytic enzymes with potential roles in cancer. *Cancer Metastasis Rev.*, **22**, 237–258.

Nielsen,H. and Krogh,A. (1998) Prediction of signal peptides and signal anchors by a hidden Markov model. *Proceedings of the 6th International Conference on Intelligent Systems for Molecular Biology*, AAAI Press, Menlo Park CA, pp. 122–130.

Nielsen,H. *et al.* (1997) A neural network method for identification of prokaryotic and eukaryotic signal peptides and prediction of their cleavage sites. *Int. J. Neural Syst.*, **8**, 581–599.

Nozaki,M. *et al.* (1999) Developmental abnormalities of glycosylphosphatidylinositol-anchor-deficient embryos revealed by Cre/LoxP system. *Lab. Invest.*, **79**, 293–299.

Sheridan,J.P. *et al.* (1997) Control of trail-induced apoptosis by a family of signaling and decoy receptors. *Science*, **277**, 818–821.

Smith,T.F. and Waterman,M.S. (1981) Identification of common molecular subsequences. *J. Mol. Biol.*, **147**, 195–197.

Stahl,N., Borchelt,D.R., Hsiao,K. and Prusiner,S.B. (1987) Scrapie prion protein contains a phosphatidylinositol glycolipid. *Cell*, **51**, 229–240.

Supek,F. and Vlahovicek,K. (2004) Inca: synonymous codon usage analysis and clustering by means of self-organizing map. *Bioinformatics*, **20**, 2329–2330.

Sutterlin,C. *et al.* (1998) *Saccharomyces cerevisiae* gpi10, the functional homologue of human pig-b, is required for glycosylphosphatidylinositol-anchor synthesis. *Biochem. J.*, **332** (Pt 1), 153–159.

Takeda,J. and Kinoshita,T. (1995) GPI-anchor biosynthesis. *Trends Biochem. Sci.*, **20**, 367–371.

Toronen,P. *et al.* (1999) Analysis of gene expression data using self-organizing maps. *FEBS Lett.*, **451**, 142–146.

Udenfriend,S. and Kodukula,K. (1995) Prediction of ω site in nascent precursor of glycosylphosphatidylinositol protein. *Methods Enzymol.*, **250**, 571–582.

Ward,J. (2001) Identification of novel families of membrane proteins from the model plant *Arabidopsis thaliana*. *Bioinformatics*, **17**, 560–563.

Chapter 4

Surface antigens and potential virulence factors from parasites detected by comparative genomics of perfect amino acid repeats

CHAPTER 4. SURFACE ANTIGENS AND POTENTIAL VIRULENCE FACTORS FROM PARASITES DETECTED BY COMPA

Surface antigens and potential virulence factors from parasites detected by comparative genomics of perfect amino acid repeats

Niklaus Fankhauser, Tien-Minh Nguyen-Ha, Joël Adler, Pascal Mäser

In press at *Proteome Science*

1 Abstract

1.1 Background

Many parasitic organisms, eukaryotes as well as bacteria, possess surface antigens with amino acid repeats. Making up the interface between host and pathogen such repetitive proteins may be virulence factors involved in immune evasion or cytoadherence. They find immunological applications in serodiagnostics and vaccine development. Here we use proteins which contain perfect repeats as a basis for comparative genomics between parasitic and free-living organisms.

1.2 Results

We have developed Reptile (http://reptile.unibe.ch), a program for proteome-wide probabilistic description of perfect repeats in proteins. Parasite proteomes exhibited a large variance regarding the proportion of repeat-containing proteins. Interestingly, there was a good correlation between the percentage of highly repetitive proteins and mean protein length in parasite proteomes, but not at all in the proteomes of free-living eukaryotes. Reptile combined with programs for the prediction of transmembrane domains and GPI-anchoring resulted in an effective tool for in silico identification of potential surface antigens and virulence factors from parasites.

1.3 Conclusions

Systemic surveys for perfect amino acid repeats allowed basic comparisons between free-living and parasitic organisms that were directly applicable to predict proteins of serological and parasitological importance. An on-line tool is available at http://genomics.unibe.ch/dora.

2 Background

Repetitive amino acid subsequences in polypeptides are of interest regarding the function as well as the evolution of proteins. At least 14% of all proteins contain internal repeats[1], the proportion being somewhat lower in prokaryote and higher in eukaryote proteomes[1]. Multicellular eukaryotes in particular, possess numerous adhesion proteins of repetitive nature in the extracellular matrix. Other highly repetitive proteins are those of the cytoskeleton[1,2]. Typical motifs involved in protein-protein interaction are the tetratricopeptide repeat (34 aa), armadillo (47 aa), ankyrin (33 aa), and the leucine-rich repeat (about 20 aa)[3].

Several tools are available for the detection of repeats in proteins: Radar (http://www.ebi.ac.uk/Radar), Repro (http://ibivu.cs.vu.nl/programs/reprowww), Internal Repeats Finder (http://nihserver.mbi.ucla.edu/Repeats), TRIPS (http://www.ncl-india.org/trips), RepSeq (http://www.repseq.gugbe.com), Rep (http://www.embl-heidelberg.de/~andrade/papers/rep/search.html), Repper (http://toolkit.tuebingen.mpg.de/repper), and ProtRepeatsDB (http://bioinfo.icgeb.res.in/repeats)[2;4–12]. Apart from simply counting repetitive occurrences of amino acid subsequences in polypeptides, repeats can be detected by self-alignment or — if the repeats are direct — by Fourier transform. Here we present Reptile, a simple tool for quantitative proteome-wide surveys of perfect amino acid repeats, and its use for the prediction of surface antigens and virulence factors from parasites.

Pathogenic bacteria as well as eukaryotic parasites often possess surface proteins of repetitive nature, presumably to protect themselves against their hosts' defence responses[13;14]. Examples are the procyclins of the sleeping sickness parasite *Trypanosoma brucei* with over twenty Glu-Pro (EP-type), respectively five Glu-Pro-Asp-Asp-Thr (GPEET-type) repeats[15;16], the circumsporozoite protein of the malaria parasite *Plasmodium falciparum* with around forty Asn-Ala-Asn-Pro (NANP) repeats[17], or SdrE from *Staphylococcus aureus*, a determinant of staphylococcal sepsis with 83 Ser-Asp (SE) repeats[18]. Such short, perfect repeats are usually very immunogenic. They may serve for serological diagnostics — the presence of repeat-directed antibodies in the serum indicating infection — as is the case with PfHRP2[19], a malaria antigen with over fifty Ala-His-His (AHH) repeats. Repetitive amino acid sequences also find applications in synthetic vaccines[20]. Furthermore, repeat-containing proteins from parasites may be virulence factors involved in immune evasion, cytoadherence, stress resistance, or biofilm formation[21–26]. The completion of the genome sequencing projects for *P. falciparum*, *T. brucei*, *Leishmania major*, and other parasites now permits systemic approaches to repeat-containing proteins.

Here we identify all proteins from pathogens that contain repeats and use them for comparative genomics between parasitic and non-parasitic species. All data and programs are freely accessible via the world-wide web.

3 Results and Discussion

3.1 Probabilistic description of perfect repeats with Reptile

In order to scan whole proteomes for repeat-containing proteins, we created the tool Reptile. It uses a "brute-force" algorithm that detects all perfect repeats and enables direct calculation of a P-value. For each input sequence, Reptile generates all possible substrings from length 2 to a user-defined maximum (the default is 20) and counts their occurrences. After removing redundant repeats that are contained within longer ones, the repeated sequences are returned by ascending P-value. The probability P to find at least n repeats of length r in a random sequence of length L (with $nr \leq L \leq n20^r$) equals the number of possible sequences that contain the desired repeat, divided by the total number of possible sequences (20^L).

$$P^*(n,r,L) = \frac{20^r 20^{L-nr}}{20^L} \binom{L-nr+n}{n} = 20^{-r(n-1)} \binom{L-n(r-1)}{n}$$

Where 20^r is the number of possible repeat sequences, 20^{L-nr} the number of possible sequences around the repeats, and the binomial equals the number of ways to place the n repeats in L. P* is an overestimate because the sequences with more than n repeats are counted too often. Taking this into account (see http://reptile.unibe.ch/pvalue.html for a more detailed description) gives the correct formula for P:

$$P(n,r,L) = \sum_{i=n}^{\frac{L}{r}} (-1)^{i+\frac{1+(-1)^{n+1}}{2}} \binom{i}{n} P^*(i,r,L)$$

Program	Method used	Detection of degenerate repeats	Calculation of a P-Value	Analysis of whole Proteomes	% Hits found in Swiss-Prot	Detection of T. brucei procyclin[1]
Reptile	Hashing[2]	No	Yes	Yes	15[3]	Yes
REP[27]	Profiles of known repeats	Yes	No	No	1.1	No
RADAR[28]	Alignment	Yes	No	No	28	Yes
REPRO[7]	Alignment	Yes	No	No	n.a.	Yes
Internal Repeats finder[8]	Alignment	Yes	Yes	No	14	No
TRIPS[9]	Fourier transform	Yes	No	No	12	No
RepSeq[10]	Hashing	Yes	Yes	Yes	n.a.	Yes
ProtRepeatsDB[11]	Mixed	Yes	Yes	Yes	n.a.	Yes
Repper[12]	Fourier transform	Yes	No	No	n.a.	No

Table 1: Comparison of programs for the detection of repetitive subsequences in proteins. [1]The T. brucei surface protein (GenBank accession AAK62893) with five GPEET repeats was used for benchmarking. [2]Word count using a hash table. [3]Using $P < 0.001$ (same as for Internal Repeats Finder).

Where i counts from n to the maximal number of repeats (L/r), switching signs with every increment according to the inclusion-exclusion principle (http://en.wikipedia.org/wiki/Inclusion-exclusion_principle). For practical purposes calculation of P*, the first summand of P, is sufficient since further summands decrease rapidly with increasing number of repeats.

Reptile returns all repeats below a user-defined cut-off P-value (the default is 10^{-5}, corresponding to an expectancy of 1 in 100000 sequences). Direct repeats are marked. The P-value being independent of the actual sequence of a repeat, Reptile also returns a measure of whether a detected repeat consists of rare or frequent amino acids. This "Amino acid abundance measure" (AM) was defined as follows:

$$AM(repeat) = log_{10}(20^r \prod_{i=1}^{r} f_i)$$

Where f_i is the frequency in the analysed proteome — respectively set of sequences submitted by the user — of the amino acid at position i of the repeat. AM is symmetric to zero, negative values indicating that a repeat predominantly consists of rare amino acids (and vice versa).

Reptile is running on-line at http://reptile.unibe.ch and accepts batch input of up to 50,000 sequences in any of the commonly used formats.

Compared to other repeat-prediction programs (Table 1) the main strengths of Reptile are its quantitative assessment of the detected repeats and its infallibility regarding short perfect repeats, such as they occur in antigens from parasites. Reptile will spot in a given protein all recurring subsequences from length two to twenty, even if they are dispersed. In contrast to programs implementing self-alignment, however, Reptile does not properly recognize degenerate repeats. Though proteins harbouring degenerate repeats also exhibit low P-values and will not go unnoticed, Reptile will not identify the basic repetitive unit but several shorter ones contained within. Other programs (Table 1) should be used when studying large repeat regions or imperfect, diverging repeats.

3.2 Genome-wide surveys for highly repetitive proteins

We defined highly repetitive proteins as proteins that contain perfect repeats of a P- value below 10^{-10}. Reptile was used to screen for such proteins highly repetitive proteins ($n \geq 3, P < 10^{-10}$)

Organism	Kingdom	Type	Proteins
Homo sapiens	Metazoa	F	38220
Mus musculus	Metazoa	F	35593
Arabidopsis thaliana	Viridiplantae	F	34554
Caenorhabditis elegans	Metazoa	F	22431
Drosophila melanogaster	Metazoa	F	16239
Brachydanio rerio	Metazoa	F	15647
Anopheles gambiae	Metazoa	F	13486
Dictyostelium discoideum	Protozoa	F	13017
Rattus norvegicus	Metazoa	F	11987
Yarrowia lipolytica	Fungi	F	6525
Saccharomyces cerevisiae	Fungi	F	5810
Kluyveromyces lactis	Fungi	F	5326
Schizosaccharomyces pombe	Fungi	F	5009
Entamoeba histolytica	Protozoa	P	9772
Giardia duodenalis	Protozoa	P	9646
Trypanosoma brucei	Protozoa	P	9210
Leishmania major	Protozoa	P	8010
Cryptococcus neoformans	Fungi	P	6569
Plasmodium falciparum	Protozoa	P	5283
Theileria parva	Protozoa	P	4071
Cryptosporidium hominis	Protozoa	P	3886
Theileria annulata	Protozoa	P	3790
Encephalitozoon cuniculi	Fungi	P	1909

Table 2: Eukaryotic proteomes analyzed. F, free-living; P, endoparasitic.

in predicted proteomes from fully sequenced genomes. The median proportion of highly repetitive proteins was 2.7% in eukaryote proteomes and 0.43% in prokaryotes, confirming the notion[1] that eukaryotes possess more repetitive proteins than bacteria ($p < 0.0001$, Mann-Whitney test). The more repeats a protein has, the longer it becomes. In eukaryotic proteomes the percentage of highly-repetitive proteins correlated to some degree with the mean protein length (Spearman coefficient $r_S = 0.51, p = 0.011$).

When distinguishing free-living from (endo)parasitic eukaryotes (Table 2), it was evident that the correlation was caused entirely by the latter. Obligate parasites exhibited a good correlation between highly-repetitive proteins and mean protein length ($r_S = 0.82, p < 0.003$) while free-living eukaryotes showed no correlation at all (Figure 1).

The finding that the percentage of highly repetitive proteins predicts average protein length only in parasite proteomes reflects the significance of repeat-containing proteins for survival in the host, possibly counterbalanced by a selective pressure on parasites for shorter proteins[29].

The eukaryote with the largest proportion of highly repetitive proteins, *Plasmodium falciparum* with 28%, and that with the smallest one, *Encephalitozoon cuniculi* with 0.42%, were both obligate parasites. The same applied to prokaryotes, where the highest proportions of highly repetitive proteins were exhibited by *Mycobacterium bovis* (3.0%), *M. tuberculosis* (2.9%) and *Parachlamydia sp.* (2.7%), and the lowest ones by *Bacillus anthracis* (Porton strain, 0.02%) and *Streptococcus pyogenes* (SSI strain, 0.05%) — however, it must be noted that with bacteria, the available genome sequences are biased towards pathogenic species.

The most repetitive protein from eukaryotes was a hypothetical protein from the sleeping sickness parasite *T. brucei*, followed by the 11-1 gene product from *P. falciparum*, a known malaria antigen of more than 1 MD size[30]. The most repetitive prokaryotic protein was a predicted cell wall surface anchor family member from *Streptococcus pneumoniae*, the leading cause of pneumonia. Table 3 summarizes these and other highly repetitive proteins identified from pathogens, emphasizing on se-

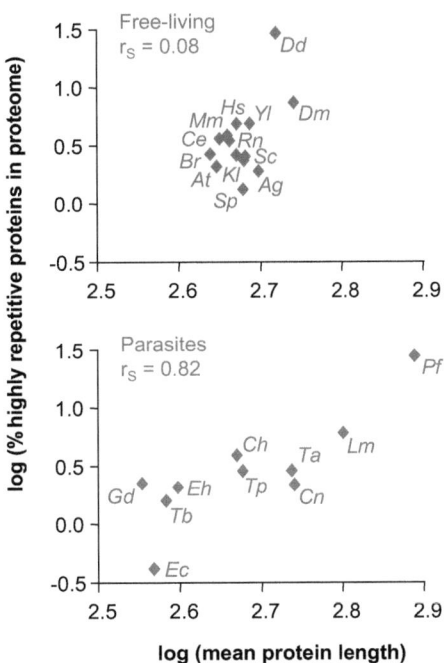

Figure 1: Comparative genomics of repeat-containing proteins. Double logarithmic plot of the percentage of highly repetitive ($n \geq 3$, $P < 10^{-10}$) proteins vs. mean protein length of eukaryotic proteomes. Ag, *A. gambiae*; At, *A. thaliana*; Br, *B. rerio*; Ce, *C. elegans*; Dd, *D. discoideum*; Cn. *C. neoformans*; Dm, *D. melanogaster*; Hs, *H. sapiens*; Kl, *K. lactis*; Mm, *M. musculus*; Rn, *R. norvegicus*; Sc, *S. cerevisiae*; Sp, *S. pombe*; Yl, *Y. lipolytica*; Ch, *C. hominis*; Cn, *C. neoformans*; Ec, *E. cuniculi*; Eh, *E. histolytica*; Gd, *G. duodenalis*; Lm, *L. major*; Pf, *P. falciparum*; Ta, *T. annulata*; Tb, *T. brucei*; Tp, *T. parva*; r_S, Spearman coefficient.

quences with experimentally verified expression.

The genome-wide surveys yielded other known virulence factors such as proteophosphoglycans of *Leishmania*[31] or PGRS (polymorphic GC-rich repetitive sequence) proteins of *Mycobacterium*, an antituberculosis vaccine candidate[32].

The presence of avirulence proteins from phytopathogenic bacteria among the most repetitive proteins indicates that repeats also serve to specifically trigger host defence responses. Remarkably repetitive are also the ice nucleation proteins of plant pathogens. Table 3 also shows examples of previously undescribed proteins. The complete datasets on repeat-containing proteins from 49 eukaryotes and 193 prokaryotes are accessible on-line via http://reptile.unibe.ch in the archive REPository.

3.3 Amino acid composition of the repeats

To further characterize the repeats, we investigated which amino acids are over- or underrepresented in repeats of $n \geq 3$, $P < 10^{-10}$ compared to the rest of the respective proteome. Overall, the amino acid composition of the repeats was more biased in eukaryotes than in bacteria (Figure 2). Small amino acids occurred more frequently in the repeats than large ones in both eukaryotes and prokaryotes. Hydrophobic residues were underrepresented in the repeats, with the exception of leucin, which in bacterial repeats was even overrepresented ($p < 0.0001$, two-tailed Wilcoxon signed rank test). Strongly overrepresented in the repeats were alanine ($p < 0.0001$) in bacteria and serine ($p = 0.0001$)

Name, accession	Sp	L	Repeat	pP
Hypothetical protein, Tb927.1.1740	Tb	7154	132x LAEESQQHTARSEADIDE	2806
Gene 11-1 protein*, Q8I6U6	Pf	10589	967x EEV	2457
Conserved protein, LmjF29.0110	Lm	3418	146xAEEQARR	1080
Proteophosphoglycan-like, LmjF35.0550	Lm	2425	105x SSSSSAPSA	1052
Putative antigen*, Tb04.29M18.750	Tb	4455	66x NEQYETLQRTNAA	958
Gb4*, Tb09.160.1200	Tb	8214	35x VVIIDCRLGSLLIDYKVI	701
Hypothetical protein, Chro.50162	Ch	1589	84x KKDAP	407
Hypothetical protein, Q8I455	Pf	2349	67x LKEEER	389
Interspersed repeat antigen*, Q8I486	Pf	1720	67x QEPVT	313
Putative antigen 332*, Q8IHN3	Pf	5507	144x EEI	274
Cell wall surface anchor family, Q97P71	Spn	4776	1074x SAS	3418
Cell surface SD repeat protein, Q88XB6	Lpl	3360	796x DS	1619
Hypothetical protein, Q8E473	Sag	1310	106x TSAS	447
Putative peptidoglycan-bound, Q8Y697	Lmo	903	78x ADADA	403
Avirulence protein, Q5GYF3	Xor	1790	20x ETVQRLLPVLCQDHGLTP	401
Serine/threonine-rich antigen, Q99QY4	Sau	2271	163x STS	391
PE-PGRS family, PG54_MYCTU	Mt	1901	136x GAG	326
Structural toxin RtxA, Q5X7A6	Lpn	7679	29x RFEDDGPVV	247
Ice nucleation protein, Q8PD38	Xca	1333	52x GYGST	242
PPE family protein, Q6MX44	Mtu	3300	95x NTG	184

Table 3: A selection of the most repetitive proteins from pathogens. Eukaryotic proteins (top) whose expression is confirmed by the presence of expressed sequence tags (EST) in GenBank are marked with an asterisk. L, length; pP, negative logarithm of the P-value; Sp, species (Ch, *C. hominis*; Lm, *L. major*; Pf, *P. falciparum*; Tb, *T. brucei*; Lmo, *Listeria monocytogenes*; Lpl, *Lactobacillus plantarum*; Lpn, *Legionella pneumophila*; Mtu, *M. tuberculosis*; Sau, *S. aureus*; Spn, *S. pneumoniae*; Sag, *Streptococcus agalactiae*; Xca, *Xanthomonas campestris*; Xor, *X. oryzae*).

in eukaryotes (Figure 2). Thus "cheap" amino acids seem to be preferred over energetically expensive ones.

Interestingly, asparagine was overrepresented in the repeats from eukaryotes but not from bacteria, suggesting that asparagines might be preferentially glycosylated in repeats. Contrary to expectation though, the probability of an asparagine to be in N-glycosylation consensus was significantly lower in repeats than in non-repetitive sequences (Figure 3). This was the case for free-living eukaryotes ($p = 0.004$) as well as for parasites ($p = 0.027$; two-tailed Wilcoxon signed rank test). The only exception was *T. brucei*, where the likelihood of an asparagine to be in N-glycosylation consensus was three-fold higher in repetitive than in non-repetitive sequences (Figure 3).

3.4 Prediction of repetitive surface antigens

In order to predict which of the repeat-containing proteins are at the cell surface, Reptile was combined with Phobius[33], a program for prediction of transmembrane domains and N-terminal export signals, and GPI-SOM[34], a program that predicts C-terminal GPI (glycosylphosphatidyl-inositol) anchor attachment sites. The three programs were run over all available proteomes predicted from completely sequenced genomes. The identified repeats were scanned for potential N-glycosylation sites. The combined output was stored in a relational database called Dora, the database of repetitive antigens, as outlined in Figure 4.

At present, Dora contains data on 1,123,238 proteins from 242 different proteomes (among which 49 eukaryotic). A www interface (http://genomics.unibe.ch/dora) allows user-defined Boolean searches. With Dora, genome-wide prediction of potential surface antigens and virulence factors is straightforward.

A search for repetitive membrane proteins in *P. falciparum* or *T. brucei* (Table 4) indeed returned

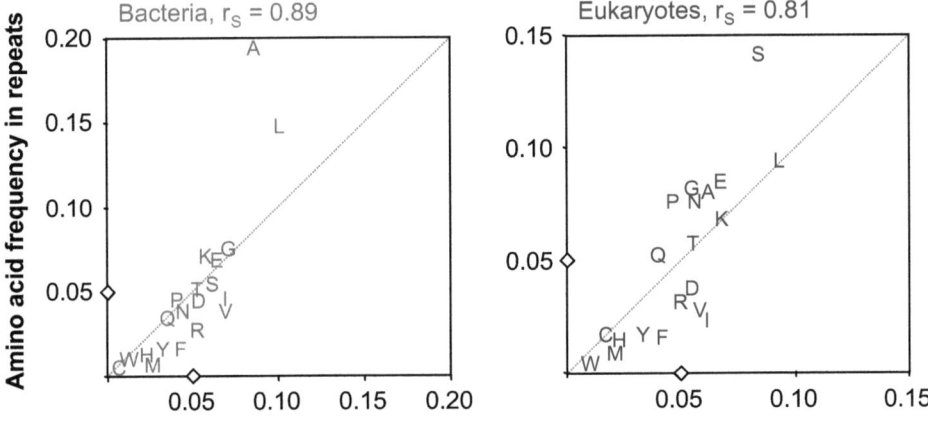

Figure 2: Amino acid composition of the repeats. For each amino acid, the frequency in the repeats of $P < 10^{-10}$ is plotted vs. its frequency in the reminders of the proteomes (r_S, Spearman coefficient). Data are pooled for bacteria ($n = 193$) and eukaryotes ($n = 49$). The small diamonds at 0.05 mark the expected frequency for random distribution, the diagonal represents equal frequency in the repeats as in the reminder of the respective proteome. Complete datatables including standard deviation are provided in the Supplement.

important surface antigens and virulence factors: circumsporozoite protein (CSP), merozoite surface proteins (MSP), erythrocyte membrane proteins (EMP), glycophorin-binding proteins (GBP), apical membrane/erythrocyte binding antigen (MAEBL), ring-infected erythrocyte surface antigen (RESA), mature parasite-infected erythrocyte surface antigen (MESA) for malaria and for *T. brucei* the procyclins, cysteine-rich acidic membrane protein (CRAM), invariant surface glycoproteins (ISG) and even the variable surface glycoproteins (VSG), which contain a significant number of dipeptide repeats (mostly AA; to our knowledge the repetitive nature of VSG was not previously recognized).

In addition to these known proteins there was a large number of uncharacterized ones, particularly from *P. falciparum* which possesses hundreds of extremely repetitive transmembrane proteins (not shown; please refer to Dora).

New specific and robust tests are urgently needed for the diagnosis of sleeping sickness, malaria, tuberculosis, and other neglected diseases[35] (see also http://www.finddiagnostics.org). PCR not being applicable in the field, serology (i.e. the detection of parasite-specific antibodies) remains the principal method of detection for many tropical diseases. Dora provides a convenient portal for identification of candidate antigens for serological tests. In addition, it can be helpful for the selection of vaccine candidates. Dora returns the hits in Fasta format, which is suitable for subsequent bioinformatic analyses.

4 Conclusions

Reptile's simple algorithm allows large-scale and quantitative description of perfect amino acid repeats. Originally designed to scan parasite proteomes for potential antigens and virulence factors, Reptile detects any protein of repetitive nature and thereby complements existing tools which work by self-alignment.

Parasite proteomes vary considerably regarding the proportion of repetitive proteins, in contrast to those of free-living eukaryotes which all contain around 3% highly repetitive ($P < 10^{-10}$) proteins.

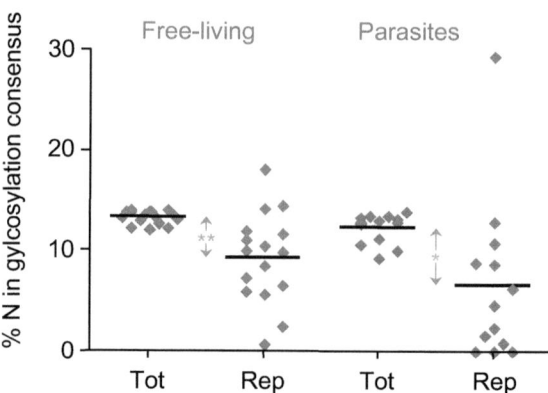

Figure 3: Potential N-glycosylation sites in the repeats. The percentage of asparagines that are in glycosylation consensus (Asp-not Pro-Ser/Thr) is plotted for repeats of $P < 10^{-10}$ and for the reminders of the respective proteomes. Bars indicate the median. The organism with 30% of asparagines in the repeats in N-glycosylation consensus is *T. brucei*.

Furthermore, the proportion of highly repetitive proteins correlates with mean protein length in parasites but not in the proteomes of free-living eukaryotes, illustrating the importance of amino acid repeats for parasites.

Scanning the predicted proteomes of parasites for amino acid repeats returned a large number of interesting proteins. Particularly useful was the combination of Reptile with prediction of glycosylation sites, export signals, transmembrane domains and GPI-anchor attachment sites, carried out on more than one million proteins from 242 different organisms.

All data are accessible on-line via Dora, database of repetitive antigens. The approach was validated against *T. brucei* and *P. falciparum*, where a Dora search returned the known surface antigens, virulence factors, and vaccine candidates plus many new, so far uncharacterized proteins.

5 Methods

5.1 Proteome files

Predicted proteome files were obtained from the Integr8 database [36] of the European Bioinformatics Institute (ftp://ftp.ebi.ac.uk/pub/databases/integr8/). The download was automated with a Python script that periodically checks for newly available proteomes, respectively for updates to previous proteome files.

5.2 Statistics

Statistical tests were performed with Prism 4.0 (GraphPad Software). Since the percentages of repeats in proteomes as well as the frequencies of amino acids were not normally distributed, non-parametric tests were used: Mann-Whitney test (http://en.wikipedia.org/wiki/Mann-Whitney_U), Wilcoxon signed rank test (http://en.wikipedia.org/wiki/Wilcoxon_signed-rank_test), and Spearman correlation (http://en.wikipedia.org/wiki/Spearman_correlation).

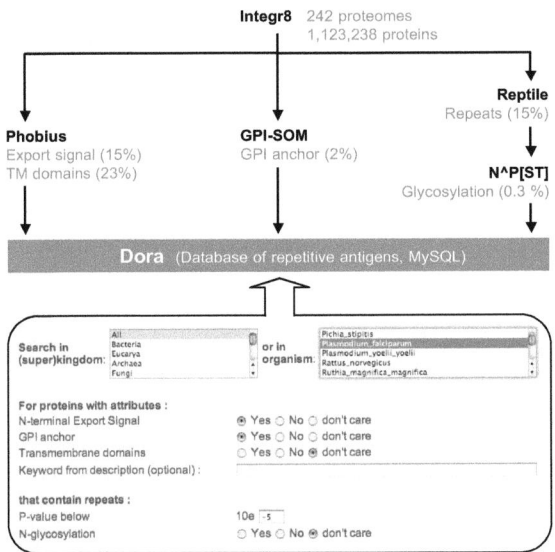

Figure 4: Flowchart to Dora, database of repetitive antigens. Reptile, Phobius[33], and GPI-SOM[34] are integrated into an automated pipeline for the classification of proteins (top). The data are stored in a database that is accessible online (http://genomics.unibe.ch/dora) via the depicted interface (bottom). This allows user-defined Boolean queries for repeat-containing surface proteins.

5.3 Reptile

The repeat detection algorithm is described under Results. The program is written in C++ and the web-interface in Perl-CGI. Reptile uses sreformat from the HMMer package[37] to convert different input formats (Fasta, GenBank, EMBL, Swiss-Prot, PIR, GCG) to Fasta. Reptile runs on a vmware (virtual infrastructure) server. The source code is available on request.

5.4 Dora

A Python script periodically runs Reptile, GPI-SOM, and Phobius over all new or updated proteome files of Integr8. . The results are stored in a MySQL database. For sake of simplicity, for each protein only the repeat with the lowest P-value is stored. A Perl script is used to interconvert Fasta format and SQL. The web interface of Dora is written in PHP. The database and all the programs run on the vmware server of the Informatics Services of the University of Bern.

6 Competing interests

The authors declare that they have no competing interests.

7 Authors' contributions

NF developed all software and generated all the data. TN designed the MySQL database and created the user interface of Dora. JA derived the formula for the calculation of the P-value. NF and PM

Name, accession	Topology	Repeat	pP
Hypothetical protein, Q8IJ50	GPI	16x EESHNFYNPTH	184
Circumsporozoite protein, Q7K740	GPI	38x ANPN	145
Merozoite surface protein 8, Q8I476	GPI	32x NN	29
Liver stage antigen, Q8IJ44	1 TM	45x AKEKLQEQQSDLEQER	839
Erythrocyte membrane protein 3, O96124	1 TM	61x QQNTGLKNTP	665
Trophozoite antigen, Q8IFL9	1 TM	60x NHKSD	287
Glycophorin-binding protein, Q8I6U8	1 TM	10x DPEGQIMREYAADPEYRKHL	213
MAEBL, Q8IHP3	1 TM	19x EEKKKADELKK	213
PF70 exoantigen, Q8IK15	3 TM	8x TKKPSKYTMNLDSPLLKGSS	165
MESA, Q8I492	1 TM	94x KE	97
PfEMP1, Q8I519	1 TM	16x GGGGS	77
RESA, Q8IHN1	1 TM	33x EEN	63
Hypothetical protein, Tb11.02.2360	GPI	11x TAVTDVNDNNSANTSNEDE	229
Hypothetical protein, Tb11.1550	GPI	12x IIAHYC	68
Procyclin (EP-type), Tb10.6k15.0020	GPI	29x PE	46
Hypothetical protein, Tb927.7.360	GPI	3x DKEKTERTEVEEVPKKDPEG	45
Procyclin (GPEET-type), Tb927.6.510	GPI	6x EETGP	24
VSG, Tb10.v4.0209	GPI	19x AA	13
CRAM, Tb10.6k15.3510	1 TM	80x ITGDCNETDDC	1050
Hypothetical protein, Tb927.3.5530	2 TM	49x RLRAEEE	337
Hypothetical protein, Tb10.61.0660	3 TM	12x NEEVPAGVSARRGGVAMSF	241
Procyclic surface glycoprotein, Tb10.26.0790	2 TM	5x YGQPPPPQ	31
Invariant surface glycoprotein, Tb927.5.350	1 TM	18x EA	12

Table 4: Repetitive membrane proteins of *P. falciparum* (top) and *T. brucei* (bottom). TM, transmembrane domain; GPI, glycosylphosphatidyl-inositol anchor; pP, negative logarithm of the P-value. See text for full protein names.

planned the study, wrote the manuscript, and designed the figures. All authors read and approved the final manuscript.

8 Acknowledgements

We wish to thank the Informatikdienste of the University of Bern for resources and support. This work was financially supported by the Swiss National Science Foundation, the Roche Research Foundation, and Biomedizin-Naturwissenschaft- Forschung Bern (TN).

References

[1] E. M. Marcotte, M. Pellegrini, T. O. Yeates, and D. Eisenberg, "A census of protein repeats," *J Mol Biol*, vol. 293, pp. 151–160, Oct 1999.

[2] M. A. Andrade, C. P. Ponting, T. J. Gibson, and P. Bork, "Homology-based method for identification of protein repeats using statistical significance estimates," *J Mol Biol*, vol. 298, pp. 521–537, May 2000.

[3] M. A. Andrade, C. Perez-Iratxeta, and C. P. Ponting, "Protein repeats: structures, functions, and evolution," *J Struct Biol*, vol. 134, pp. 117–131, May 2001.

[4] R. Szklarczyk and J. Heringa, "Tracking repeats using significance and transitivity," *Bioinformatics*, vol. 20 Suppl 1, pp. 311–317, Aug 2004.

[5] A. Heger and L. Holm, "Rapid automatic detection and alignment of repeats in protein sequences," *Proteins*, vol. 41, pp. 224–237, Nov 2000.

[6] K. B. Murray, W. R. Taylor, and J. M. Thornton, "Toward the detection and validation of repeats in protein structure," *Proteins*, vol. 57, pp. 365–380, Nov 2004.

[7] R. A. George and J. Heringa, "The repro server: finding protein internal sequence repeats through the web.," *Trends Biochem Sci*, vol. 25, pp. 515–517, October 2000.

[8] M. Pellegrini, E. M. Marcotte, and T. O. Yeates, "A fast algorithm for genome-wide analysis of proteins with repeated sequences," *Proteins*, vol. 35, pp. 440–446, Jun 1999.

[9] M. V. Katti, R. Sami-Subbu, P. K. Ranjekar, and V. S. Gupta, "Amino acid repeat patterns in protein sequences: their diversity and structural-functional implications," *Protein Sci*, vol. 9, pp. 1203–1209, Jun 2000.

[10] D. P. Depledge, R. P. J. Lower, and D. F. Smith, "Repseq - a database of amino acid repeats present in lower eukaryotic pathogens," *BMC Bioinformatics*, vol. 8, pp. 122+, April 2007.

[11] M. K. Kalita, G. Ramasamy, S. Duraisamy, V. S. Chauhan, and D. Gupta, "Protrepeatsdb: a database of amino acid repeats in genomes," *BMC Bioinformatics*, vol. 7, pp. 336+, July 2006.

[12] M. Gruber, J. Soding, and A. N. Lupas, "REPPER–repeats and their periodicities in fibrous proteins," *Nucleic Acids Res*, vol. 33, pp. 239–243, Jul 2005.

[13] R. W. Leid, C. M. Suquet, and L. Tanigoshi, "Parasite defense mechanisms for evasion of host attack; a review," *Vet Parasitol*, vol. 25, pp. 147–162, Jul 1987.

[14] L. Kedzierski, J. Montgomery, J. Curtis, and E. Handman, "Leucine-rich repeats in host-pathogen interactions," *Arch Immunol Ther Exp (Warsz)*, vol. 52, pp. 104–112, Mar 2004.

[15] I. Roditi, M. Carrington, and M. Turner, "Expression of a polypeptide containing a dipeptide repeat is confined to the insect stage of Trypanosoma brucei," *Nature*, vol. 325, pp. 272–274, Jan 1987.

[16] E. Vassella, A. Acosta-Serrano, E. Studer, S. H. Lee, P. T. Englund, and I. Roditi, "Multiple procyclin isoforms are expressed differentially during the development of insect forms of Trypanosoma brucei," *J Mol Biol*, vol. 312, pp. 597–607, Sep 2001.

[17] V. Enea, J. Ellis, F. Zavala, D. E. Arnot, A. Asavanich, A. Masuda, I. Quakyi, and R. S. Nussenzweig, "DNA cloning of Plasmodium falciparum circumsporozoite gene: amino acid sequence of repetitive epitope," *Science*, vol. 225, pp. 628–630, Aug 1984.

[18] S. J. Peacock, C. E. Moore, A. Justice, M. Kantzanou, L. Story, K. Mackie, G. O'Neill, and N. P. J. Day, "Virulent combinations of adhesin and toxin genes in natural populations of Staphylococcus aureus," *Infect Immun*, vol. 70, pp. 4987–4996, Sep 2002.

[19] C. Beadle, G. W. Long, W. R. Weiss, P. D. McElroy, S. M. Maret, A. J. Oloo, and S. L. Hoffman, "Diagnosis of malaria by detection of Plasmodium falciparum HRP-2 antigen with a rapid dipstick antigen-capture assay," *Lancet*, vol. 343, pp. 564–568, Mar 1994.

[20] G. Snounou and L. Renia, "The vaccine is dead–long live the vaccine," *Trends Parasitol*, vol. 23, pp. 129–132, Apr 2007.

[21] S. Brinster, B. Posteraro, H. Bierne, A. Alberti, S. Makhzami, M. Sanguinetti, and P. Serror, "Enterococcal leucine-rich repeat-containing protein involved in virulence and host inflammatory response," *Infect Immun*, vol. 75, pp. 4463–4471, Sep 2007.

[22] U. Samen, B. Eikmanns, D. Reinscheid, and F. Borges, "The surface protein Srr-1 of Streptococcus agalactiae binds human keratin 4 and promotes adherence to epithelial HEp-2 cells," *Infect Immun*, Aug 2007.

[23] F. M. Tomley, K. J. Billington, J. M. Bumstead, J. D. Clark, and P. Monaghan, "EtMIC4: a microneme protein from Eimeria tenella that contains tandem arrays of epidermal growth factor-like repeats and thrombospondin type-I repeats," *Int J Parasitol*, vol. 31, pp. 1303–1310, Oct 2001.

[24] J. de la Fuente, J. C. Garcia-Garcia, A. F. Barbet, E. F. Blouin, and K. M. Kocan, "Adhesion of outer membrane proteins containing tandem repeats of Anaplasma and Ehrlichia species (Rickettsiales: Anaplasmataceae) to tick cells," *Vet Microbiol*, vol. 98, pp. 313–322, Mar 2004.

[25] F. Ansari, N. Kumar, M. Bala Subramanyam, M. Gnanamani, and S. Ramachandran, "MAAP: Malarial adhesins and adhesin-like proteins predictor," *Proteins*, Aug 2007.

[26] I. Cherny, L. Rockah, O. Levy-Nissenbaum, U. Gophna, E. Z. Ron, and E. Gazit, "The formation of Escherichia coli curli amyloid fibrils is mediated by prion-like peptide repeats," *J Mol Biol*, vol. 352, pp. 245–252, Sep 2005.

[27] M. A. Andrade, C. P. Ponting, T. J. Gibson, and P. Bork, "Homology-based method for identification of protein repeats using statistical significance estimates," *J Mol Biol*, vol. 298, pp. 521–537, May 2000.

[28] A. Heger and L. Holm, "Rapid automatic detection and alignment of repeats in protein sequences," *Proteins*, vol. 41, pp. 224–237, Nov 2000.

[29] M. D. Katinka, S. Duprat, E. Cornillot, G. Metenier, F. Thomarat, G. Prensier, V. Barbe, E. Peyretaillade, P. Brottier, P. Wincker, F. Delbac, H. El Alaoui, P. Peyret, W. Saurin, M. Gouy, J. Weissenbach, and C. P. Vivares, "Genome sequence and gene compaction of the eukaryote parasite Encephalitozoon cuniculi," *Nature*, vol. 414, pp. 450–453, Nov 2001.

[30] C. Petersen, R. Nelson, J. Leech, J. Jensen, W. Wollish, and A. Scherf, "The gene product of the Plasmodium falciparum 11.1 locus is a protein larger than one megadalton," *Mol Biochem Parasitol*, vol. 42, pp. 189–195, Sep 1990.

[31] T. Ilg, "Proteophosphoglycans of Leishmania," *Parasitol Today*, vol. 16, pp. 489–497, Nov 2000.

[32] J. Campuzano, D. Aguilar, K. Arriaga, J. C. Leon, L. P. Salas-Rangel, J. Gonzalez-y Merchand, R. Hernandez-Pando, and C. Espitia, "The PGRS domain of Mycobacterium tuberculosis PE_PGRS Rv1759c antigen is an efficient subunit vaccine to prevent reactivation in a murine model of chronic tuberculosis," *Vaccine*, vol. 25, pp. 3722–3729, May 2007.

[33] L. Kall, A. Krogh, and E. L. L. Sonnhammer, "A combined transmembrane topology and signal peptide prediction method," *J Mol Biol*, vol. 338, pp. 1027–1036, May 2004.

[34] N. Fankhauser and P. Maser, "Identification of GPI anchor attachment signals by a Kohonen self-organizing map," *Bioinformatics*, vol. 21, pp. 1846–1852, May 2005.

[35] M. Usdin, M. Guillerm, and P. Chirac, "Neglected tests for neglected patients," *Nature*, vol. 441, pp. 283–284, May 2006.

[36] M. Pruess, P. Kersey, and R. Apweiler, "The Integr8 project–a resource for genomic and proteomic data," *In Silico Biol*, vol. 5, no. 2, pp. 179–185, 2005.

[37] S. R. Eddy, "Multiple alignment using hidden Markov models," *Proc Int Conf Intell Syst Mol Biol*, vol. 3, pp. 114–120, 1995. Comparative Study.

CHAPTER 4. SURFACE ANTIGENS AND POTENTIAL VIRULENCE FACTORS FROM PARASITES DETECTED BY COMP.

Chapter 5

Comparative transportomics between parasites and free-living eukaryotes

Comparative transportomics between parasites and free-living eukaryotes

Niklaus Fankhauser and Pascal Maeser

Working paper

1 Abstract

1.1 Background

The aim of this project is to identify differences between parasites and free-living eukaryotes, assign possible function to unknown protein families and learn more about the role of protein families in evolution. It makes use of the large number of completely sequenced genomes of unicellular eukaryotic parasitic and free-living organisms that have become available in recent years. We are interested in the transmembrane protein subset of these organisms, because in this zone of interaction significant differences can be found.

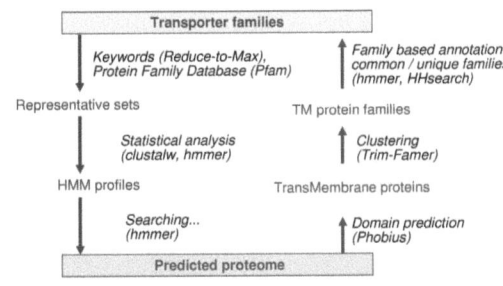

Figure 1: Schematic overview of the Top-Down and Bottom-Up approach.

1.2 Results

In some aspects of their transportome, parasites and free-living cells are almost indistinguishable. In others, we can clearly observe them to form separate groups. This study gives an overview on what can be learned about the difference between parasites and free living organisms by looking at their sets of transporters.

1.3 Conclusions

Parasites have a smaller fraction of transmembrane proteins used as transporters, but this difference stems only from certain subsets of transporters. TransOrgalin expands our understanding of hyper-dimensional protein-family clusters, as it provides us with the ability to move though them and watch their most interesting aspects from exciting viewpoints. It is available at http://transorgalin.unibe.ch.

2 Background

Much effort has been done to discover the cunning strategies of protozoan parasites, who are responsible for most severe diseases in the tropical world. Those parasites include *Trypanosoma brucei*, which causes sleeping sickness, *T. cruzi*, which causes Chagas disease, *Plasmodium* species, which cause malaria, *Theileria parva*, a tick-borne parasite that causes East Coast fever in cattle, *Toxoplasma gondii*, a protozoan parasite that causes birth defects and attacks AIDS patients as well as *Entamoeba histolytica*, which causes amoebic dysentery.

Compared to bacterial infections, these parasitemias are much harder to treat, often requiring drugs with severe side effects for the patient. Because humans are much closer related to these parasites, targeting their essential enzymes also affects our own cells. This made it necessary to learn more about the intricate host-parasite interaction by sequencing their genomes. Many eukaryotic genomes, of parasites as well as free-living organisms, have become available in the past years.

Our main hypothesis for further investigation of eukaryotic parasites is, that they require specialised transporter proteins to salvage nutrients from their hosts. It is therefore comparing the sets of transporters, the transportome, of parasites and free-living eukaryotes that should show important differences.

The fact that there are already many excellent databases of protein families publicly available made it possible to perform a top-down approach, where we look at the differences in known family composition of eukaryotic organisms. As shown in the left part of Figure 1, we start from transporter families, which we find by keyword search in sequence annotations or in a protein family database. Analysis of the search results for the family profiles thus acquired should reveal differences between parasites and free-living organisms. Our hypothesis is that parasites have a relatively larger transportome than free-living organisms.

Clustering proteins into families of similar sequences, the bottom-up approach, is illustrated in the right part of Figure 1. We know that protein families exist in all organisms, as they were created by common evolutionary duplication events. They can also be the source of redundancy of parasitic drug-resistance. We are looking for conserved protein families in parasites (preferably transporters) which have no counterparts in the human proteome.

3 Methods

3.1 Data sets

The starting point of the Top-Down approach was a set of hidden Markov model (HMM) profiles[1] describing experimentally verified transporter proteins. Additionally, five sub-sets of profiles for ion, ABC (ATP-binding cassette)[2], sugar, amino acid and nucleoside/tide transporters were used.

The set of transporter proteins was created by reading through descriptions in the Pfam[3] database and choosing the 122 profiles describing transporter proteins. Of those profiles, 99 are actually used by eukaryotes, the rest are exclusively used by the other two superkingdoms. The set of ion transporter HMMs was created by searching for protein sequences of transporters for magnesium, potassium, calcium and zinc ions. The remaining data sets are selected Pfam profiles found by keyword searches.

The 25 proteomes of free-living organisms used are from *Anopheles gambiae, Arabidopsis thaliana, Ashbya gossypii, Aspergillus fumigatus, Aspergillus oryzae, Brachydanio rerio, Caenorhabditis briggsae, Caenorhabditis elegans, Candida glabrata, Chlamydomonas reinhardtii, Debaryomyces hansenii, Dictyostelium discoideum, Drosophila melanogaster, Drosophila pseudoobscura, Gibberella zeae, Homo sapiens, Kluyveromyces lactis, Mus musculus, Neurospora crassa, Oryza sativa, Rattus norvegicus, Saccharomyces cerevisiae, Schizosaccharomyces pombe, Tetraodon nigroviridis* and *Yarrowia lipolytica*.

The 14 parasites considered are *Cryptococcus neoformans, Cryptosporidium hominis, Encephalitozoon cuniculi, Entamoeba histolytica, Giardia lamblia, Leishmania major, Plasmodium falciparum, Plasmodium yoelii yoelii, Theileria annulata, Theileria parva, Toxoplasma gondii, Trypanosoma brucei, Trypanosoma cruzi* and *Ustilago maydis*.

3.2 Data preparation

To create the ion transporter HMM profiles, the Hmmer package[4] was used on the clustalw[5] multiple alignments of the sequences. The set of protein sequences was first cleared from redundant sequences by a self-written perl program called "Reduce-To-Max", using Needleman-Wunsch alignment[6] scores to identify similar sequences. The above described HMM profiles were then searched against all proteins containing more than one predicted transmembrane domain from those 39 organisms. These computations, performed on an AMD64 computer running gentoo linux, took a few days to complete. The result of these searches (protein-ID, profile-ID, P-value) are stored in a MySQL database for further analysis. The process is summarised in the left part of Figure 1. On these results, the following types of analysis were then performed.

3.3 Mean-value distribution analysis

Mean values for the five profile sets were calculated in the Python programming language from the values stored the the database. A Mann-Whitney U test with the null hypothesis stating that the mean values are the same in the parasite as in the free-living group was performed using SciPy[7] library. Graphical data displays were produced from the resulting comparison using the Ploticus software (ploticus.sf.net).

3.4 Kohonen self organising maps

SOM analysis[8] was performed using the SprAnnLib C library[9], which we previously used successfully to identify the GPI anchoring signal[10] in the C-terminus of protein sequences. Using the Simplified Wrapper and Interface Generator SWIG[11], a python module was created from the C function that creates and runs the neural network. A graphical user interface was implemented in wxPython to simplify fine-tuning of network parameters and visualisation of labelled SOMs. There, the data for the input vectors is acquired from the database and the P-value threshold as well as the normalisation method can be specified. After training of the network, which takes about 2 minutes for a 10x15 neuron SOM on a Intel CoreDuo, the winning neurons can be observed for al training cycles. Neurons are optimally labelled in real-time with the organisms activating them. This makes it possible to see how a SOM gradually improves categorisation of the input data with increasing number of training cycles and to determine after how many cycles the network eventually becomes over-trained. The optimal size for a SOM providing a good separation depends on the number of organisms analysed and was determined empirically.

3.5 Principal components analysis

The 'prcomp' function of the R Project for Statistical Computing[12] was used to determine the principal components of the Pfam transporter set. Each row of Pfam hits was normalised by the maximum hit count in that row. The original data was then expressed in terms of only the two most significant component using the 'biplot' function. In other words, the dimensionally reduced data was plotted bases on the two orthogonal eigenvectors with the largest eigenvalues.

3.6 Average-linkage clustering

Average-linkage clustering using Kendall's tau as distance measure was performed using the command-line version of Cluster 3.0 by Michiel de Hoon of the University of Tokyo, based on the algorithm by Michael Eisen[13]. The normalised number of hits per proteome below the P-value threshold of 10^{-10} was used as input data. The clustered data was then visualised by a self-written Python program to display trees circularly and by Java TreeView[14].

3.7 Transportome clustering

In the bottom-up approach, we used an organism subset consisting of parasites and *H. sapiens*, as listed in Table 1. The following calculations were performed for each organism separately to create the protein family HMM profiles, which were then searched against those of all other organisms, as illustrated in the right part of Figure 1. The starting-point for the process of creating protein families is the matrix of Needleman-Wunsch alignment scores of every transmembrane protein against all others in the transportome. These and all following calculations, which would consume an enormous amount of time if performed on an standard desktop PC, were done on the University of Bern Linux Cluster (ubelix). The transmembrane domains were predicted using the Phobius combined transmembrane topology and signal peptide predictor[15] and stored in the database. All sequences with more than one predicted transmembrane domains were used as the transportome.

The clustering algorithm, written in Python, was inspired by a program used to create protein families in plants[16], but uses an adaptive threshold based on statistical properties of each protein family. Initially, the program called "Trim-Famer" clusters proteins into families if they share more than 15% similarity. This threshold is incremented until the number of families is maximised while the families are trimmed multiple times at each step. In order to trim a family, the mean value of all its protein similarity scores is determined and scores below 10% of the mean above the current threshold are removed from the matrix. The resulting matrix is then used to re-cluster the proteins. Fifty trim-steps are evaluated when optimising the number of families. This method reveals families sharing a common functional domain, like an ATP-binding site, as separate entities, when otherwise weak links would have joined them artificially.

3.8 Family analysis

The protein sequences thus grouped as families are now aligned using clustalw and visualised scaled to constant width. The alignment colours, based on similarity and chemical properties, were assigned by mview [17] and the positions of their transmembrane domains plotted in relation to the alignment. Hidden Markov models were generated by Hmmer and then searched against each others using HHsearch [18]. Numerical results are stored in the database and images on the web-server, where they are available though the TransOrgalin interface. This allows the following overviews and in detail information:
First a table of all families for each organism with summary information on each family. It is possible to use filters and perform keywords searches. Families are annotated automatically by choosing the most reliable name of the family. By taking into account HMM-HMM search hits, more annotation for families that have no information in another organism can be derived.
There is a page summarising these results statistically, showing organism properties like the size of the transportome, number of families created and percentages of different family types.
For each each protein family, there is a detailed view with sequence names, dendrogram, multiple alignment, transmembrane domains and HMM-HMM search hits. The same analysis is also available for pairs of similar protein families and their pooled sequences.
The occurrence of protein families in different organisms can be seen in a tabular representation. It is possible to search iteratively for connected protein families of one protein family. The results of iterative keyword searches can be clustered and displayed as linear or circular dendrograms. Relations between selectable groups of organisms can be studied by looking at common protein families.

4 Results and Discussion

4.1 Transmembrane/transporter fractions

In the top-down approach, we are looking for characteristic differences in the occurrence of well-known protein families in parasitic eukaryotes. The first plot in Figure 2 compares the fraction of transmembrane proteins in parasites and free-living organisms. According to the Mann-Whitney U test, the null hypothesis cannot be rejected; the two groups are indistinguishable by their percentage of transmembrane proteins. The comparison below shows the mean percentage of Pfam transporter profile hits to be clearly lower in parasites than in free-living organisms. Our working hypothesis has been that parasites should have a larger fraction of transmembrane proteins used as transporters than free-living organisms. It is based on the fact that the percentage of transmembrane proteins is near identical in the two groups, but parasites were thought to rely more heavily on transporters to compensate for deficits of their reduced metabolisms. As this hypothesis is rejected, the question remains why there are fewer transporters found in parasites. One possible explanation is that parasites have a significant number of unknown transporters, as they are less extensively explored as organisms more closely related to *H. sapiens*. This possibility is elaborated in Section 4.7.

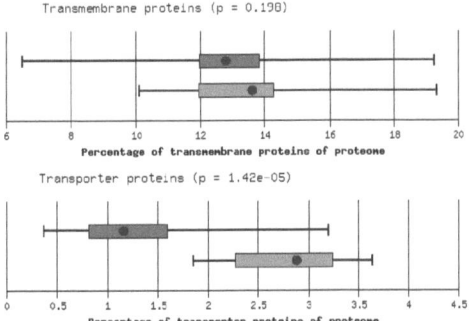

Figure 2: Box-plot analysis of mean value differences in parasites (red) and free-living organisms (green) transmembrane protein group percentages. The number in parenthesis after the title of each plot is the p-value of a Mann-Whitney U test with the null hypothesis stating that the mean values are the same in the parasite as in the free-living group. Dot: median. Box: 25th through 75th percentile. Tails: 5th and 95th percentile.

It is also possible that parasites have a larger fraction of one group of transporters, compensated by a smaller fraction of an other group. This question will be addressed in Section 4.5.

4.2 Self Organising Map

The SOM analysis (Figure 4) of the transporter protein family set shows that parasites and free living eukaryotes form distinct groups on the map. Organisms very similar in respect to their transportome activate the same neuron on the map, like for example most animals. But no single neuron responds to both parasites and free-living organisms.
Clustered at the top of the map, free living organisms appear as the three expected groups of fungi, plants and animals. Parasites cover the lower half of the map. They are mostly activating distinct neurons, with the exception of the kinetoplastidae and and a group consisting of *E. cuniculi, E. histolytica, T. annulata* and *T. parva*. The two free-living organisms positioned most closely to the parasite cluster are *D. discoideum* and *C. reinhardtii*, but they are still separated. These two organisms are evolutionary most closely to the parasites. The parasitic fungi *U. maydis* marks the left extreme of the parasite cluster and is positioned towards the fungi cluster. A large group of animals and some fungi clustered at one neuron seem to suggest that those different organism groups have a more similar transportome than the flies *A. gambiae, D. melanogaster* and *D. pseudoobscura*, which are clustered to the left of the other group.
This analysis confirms that parasites and free-living organisms can actually be distinguished by their transportome profile. The SOM approach is a flexible way to explore the structure of such data without prior knowledge about it. Our graphical user interface to the neural network made it easy to try different network topologies and transformations of the input data.

4.3 Principal Component Analysis

The two most significant components, which were used for the plot, contribute 52% to the total variance. As in the SOM analysis, the organisms are clustered in the groups of parasites, animals and

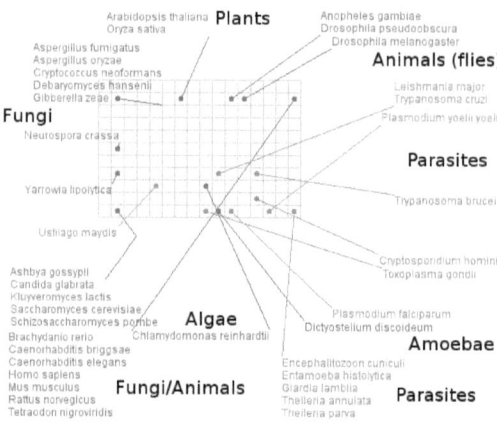

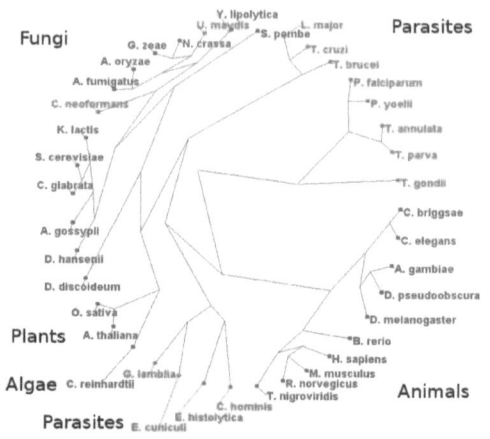

Figure 3: 10x15 SOM of search results from 80 transporter HMM in 39 eukaryotic transporter proteomes. A p-value threshold of 10^{-10} was used and the network was trained for 1100 cycles. Each square of the map represents a neuron. Responding neurons are marked by dots labelled with the names of the organisms activating them. Dots and label are coloured according to whether the input vector was from a parasite (red) or a free-living organism (green). The following organism groups are indicated next to their members' names: animals, fungi, algae, amoebae and parasites.

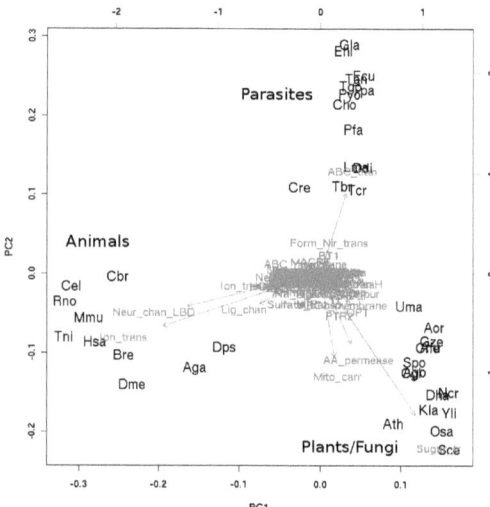

Figure 4: R 'biplot' of transporter data set reduced in dimensionality by PCA and plotted in terms of the two most significant components. Organism names (black) are abbreviated to three letters. The names of the Pfam profiles (red) are next to the arrows representing their axes.

Figure 5: Circular dendrogram of eukaryotic organisms based on the number of hits of a set of transporter Pfam profiles. Labels in red are parasites and those in green are free-living organisms. The root of the tree is in the centre of the ellipse and branches outward. The same organism groups as in Figure 2 are indicated next to the ellipse segment of its members.

fungi. The Pfam profiles contributing most to this separation can be seen in red next to their cluster. They are listed in Section 4.4.

4.4 Average-linkage clustering

The transporter family set search results were average-linkage clustered and displayed as circular tree in Figure 5, to get an overview of this alternative clustering. The separation of parasites and free-living organisms is also evident. There are three distinct, large groups of parasites and two outsider parasites. One group consists of the kinetoplastidae (*T. brucei, T. cruzi, L. major*). The second consists of the plasmodiae (*P. falciparum, P. yoelii*), the theileriae (*T. parva, T. annulata*) and *Toxoplasma gondii*. These two groups are most closely grouped to the fungi and animal cluster. The final large parasite group consists of *G. lamblia, E. cuniculi, E. histolytica* and *C. hominis*. It is grouped next to the plant/algae and the animal cluster. The two outsiders are the parasitic fungi *U. maydis* and *C. neoformans*, which are grouped among the free-living fungi.

Figure 6 shows this clustered data displayed by Java Tree-View to visualise how the input data directs the clustering. Pfam profiles distinguishing groups of parasites can be recognised here, like for example ABC transporter, (ABC_tran), biopterin transporter (BT), membrane-attack-complex perforin (MACPF), formate-nitrite transporter (Form_Nir_trans) and phosphate transporter (PHO4).

4.5 Distribution in sub-sets

To get a better understanding of how exactly the groups are separated, statistical mean value comparison of parasites and free-living organisms was done for different sub-sets of transporter protein profiles. Box-plots of these comparisons can be seen in Figure 7. Interpretation of the significance tests leads to the conclusion that the average fraction of Pfam transporter profile hits is lower in parasites than in free-living organisms for the ion, sugar and with a slightly less confident P-value for the amino acid transporter subset. The ABC and nucleobase transporter groups on the other hand show no significant mean value differences. The most successful pattern profile group for discerning parasites is the ion transporter set. This leads to the conclusion that these proteins should be further investigated in parasites. Possibly they have a smaller fraction of ion transporters because they do not need such a broad range of them, as they can move to location in their host providing favourable ion concentrations. Alternatively, it could also be lower because there is a significant number of completely unknown ion transporter proteins in parasites.

4.6 Transportome clustering

For each organism in the analysis (Table 1), between 18 (*P. yoelii*) and 330 (*H. sapiens*) protein families were created in the bottom-up clustering process. Information on these families like annotation, sequences, phylogenetic trees, transmembrane domain predictions and related families can be conveniently accessed by the TransOrgalin web-interface. Table 1 provides statistical properties of the resulting data.

In the first column, *H. sapiens* has the highest number of proteins clustered in families, which was to be expected for a complex multi-cellular organism. The parasites on the other hand vary in a wide range of familome percentages. *E. cuniculi*, a parasites with a very small proteomes, has a surprisingly high percentage of proteins clustered into families.

The next column ("unique") shows the percentages of protein families that have no HMM-HMM search results in any of the organisms in this analysis above the threshold of 10^{-10}. Besides *H. sapiens*, the top quarter of unique family percentages contains the two parasites *T. parva* and *G. lamblia*. These two are apparently isolated from the bulk of sequenced parasites and each other by higher evolutionary specialisation. They are also the species with the lowest fraction of transmembrane proteins organised in families.

The last column indicates the percentage of families without HMM-HMM hits in human protein families. With the exception of *C. hominis* as well as the kinetoplastidae *L. major* and *T. brucei*, all parasites show a high percentage of families without human analogs. Interestingly, the third kinetoplastidae in our analysis, *T. cruzi*, has a much higher non-human percentage than his two close evolutionary relatives.

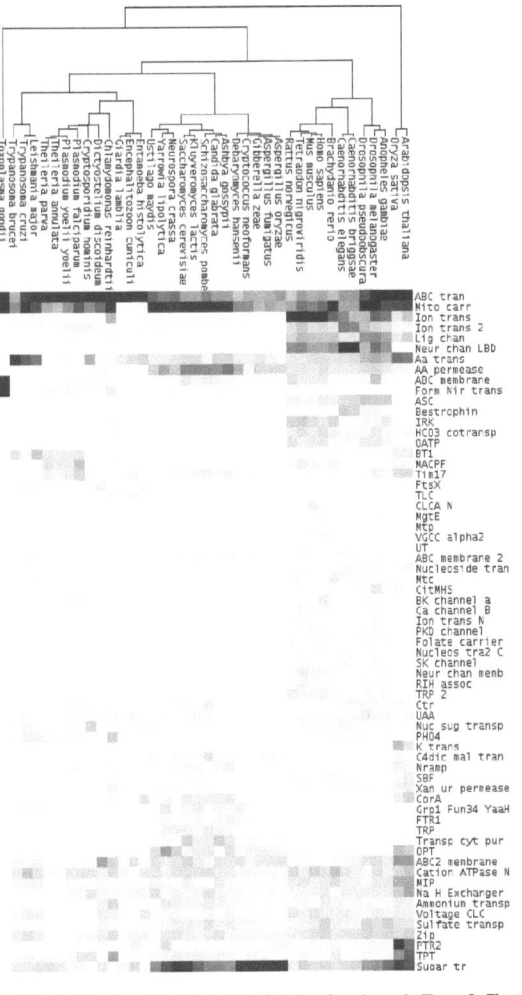

Figure 6: Java TreeView visualisation of the same clustering as in Figure 5. The input data resulting in the observed clustering is visible. On the horizontal axis are the organisms, while the names of the Pfam profile are on the vertical axis. The profiles were also clustered. Positive values are shades of black.

Organism	Familome	Unique	Nonhuman
Homo sapiens	7.65	86.36	0
Trypanosoma cruzi	5.71	69.49	98.31
Plasmodium yoelii yoelii	4.76	50	88.89
Encephalitozoon cuniculi	4.56	65	80
Entamoeba histolytica	3.87	70.42	88.73
Cryptosporidium hominis	3.29	58.33	62.5
Plasmodium falciparum	3.12	50	95.45
Leishmania major	3.03	26.67	57.78
Trypanosoma brucei	2.89	32.56	53.49
Theileria parva	2.78	80.95	100
Giardia lamblia	1.52	79.31	89.66

Table 1: Selected properties of eukaryotic proteomes. The familome is the percentage of transmembrane proteins clustered into families. Unique is the percentage of protein families that have no HMM-HMM hit in any other organism. Non-human is the percentage of families with no hits to *H. sapiens* protein families. The list is sorted by largest percentage of familome first.

Family	T. brucei	T. cruzi	L. major	H. sapiens
ABC transporter	12	104	12	227
Cation transporter	15	13	6	212
Aquaporin	17	103	4	315
Amino acid tp.	23	100	43	256
Phospholipid tp.	26	58	19	185
Protein kinase	29	60	8	
Adenylylcyclase	38	27	16	

Table 2: Part of the table representing protein family occurrences in the TransOrgalin web interface sorted by T. brucei, T. cruzi, L. mayor, followed by H. sapiens. The first column are the annotations of protein families. The following columns, one for each organism, shows whether the family in this row occurs in an organism. If the family is present in a table cell, it contains a TransOrgalin family ID number.

4.7 Family analysis

Special care was taken to meaningfully annotate the protein families. If the sequences of a protein family contain multiple annotations including some containing keyword like "unknown" and "hypothetical", which are not considered, then the most common annotation is taken for the whole family. Should there be nothing but "hypothetical" annotations, the search for a functional description is extended to similar protein families in different organisms, as indicated by HMM-HMM results. In this way, functional annotations could be added in most analysed organisms.

An average of 12% of protein families in parasites had their function only revealed by using this sensitive method. In contrast, less than 1% of families in H. sapiens could be newly annotated. This supports the hypothesis that parsite genomes contain unknown transporters. These transporter families were not detected by available profiles. The effective percentage of transporters in Figure 2 is therefore increased, as discussed in Section 4.1.

The interface can also be used to explore how a certain protein family is differently used by each parasite. Starting for example with a keyword search for multidrug resistance proteins, four proteins families from four different parasites (G. lamblia, T. cruzi, T. brucei, L. mayor) contain the "multidrug" keyword. By recursively following all HMM-HMM search results, 65 additional protein families from all organisms in the analysis are found. A circular dendrogram is constructed where connections between two leafs of the tree represent HMM-HMM relations.

Because we are interested in exploring new families of pharmacological exploitable transmembrane proteins, retrieving protein families occurring in some organisms while being absent in others is essential. An possible application of this feature of TransOrgalin is shown in Table 2. A list of protein families is assembled by using a hierarchy of organisms as clustering centres to show in which organism a family occurs. Setting the three kinetoplastidae parasites as the highest priority followed by H. sapiens shows that two protein families occur only in the parasites, but not in H. sapiens. The first is an adenylate cyclase family that has already been studied[19] and the second is an unexplored family of protein kinases that could be further investigated experimentally. The SwissProt IDs of that protein family are Q382D0, Q57VH8, Q38C51 for T. brucei, Q4DN46, Q4DUZ9, Q4DY37, Q4DY39 for T. cruzi, and Q4Q147, Q4Q920, Q4Q6Y5 for L. major.

The opposite view of looking at all protein families that are common in a selected set of organisms can also be clustered and visualised as a circular or linear tree. It is also possible to automatically cut the tree into branches in order to present the common protein families in an organised way.

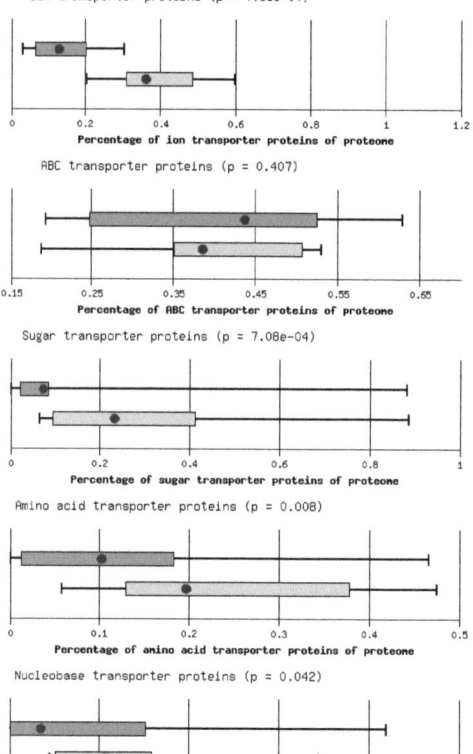

Figure 7: Box-plot analysis of mean value differences in parasites (red) and free-living organisms (green) transmembrane protein group percentages. The number in parenthesis after the title of each plot is the p-value of a Mann-Whitney U test with the null hypothesis stating that the mean values are the same in the parasite as in the free-living group. Dot: median. Box: 25th through 75th percentile. Tails: 5th and 95th percentile.

5 Conclusions

Contrary to our initial hypothesis, parasites use a smaller fraction of their transmembrane proteins as transporters. The idea that parasites have a larger fraction of one group of transporters, compensated by a smaller fraction of an other group also has to be revised. The transporter sub-sets in our analysis had equal percentages in both parasites and free-living organisms, except for the ion, sugar and amino acids transporters, which show a significantly lower percentage in parasites.

The self organising map analysis of the transportome profiles shows that this pattern is sufficient to distinguish between parasite and free-living organisms, as no neuron responds to both of them. *D. discoideum* and *C. reinhardtii* were closely clustered to the parasites because they are most similar to them. This separation of parasites could be confirmed by principal component analysis as well as average-linkage clustering, where it becomes apparent that the two parasitic fungi *U. maydis* and *C. neoformans* are harder to distinguish from their free-living fungal relatives. Three transporter (biopterin, formate-nitrite, phosphate) and a perforin Pfam profile could be identified as contributing most to the separation of parasites in the clustering.

Creating transmembrane protein families de novo in parasites and *H. sapiens* made even more of their peculiarities become apparent. *T. parva* and *G. lamblia*, the two parasites with the smallest transmembrane fraction also turned out to be the ones with the most families without any significant similarities to other organisms' families. The results of the family interactions established by HMM-HMM matches and the family annotations improved by them is fully available to the research community through the TransOrgalin interface. Using it to find families only present in the kinetoplastidae parasites revealed that they exclusively share a family of adenylate cyclases and protein kinases not present in *H. sapiens*.

6 Acknowledgements

We wish to thank the members of the groups of Thomas Seebeck, Ernst Schweingruber, Isabel Roditi and Beatrice Lanzrein for their helpful advice and Marc Mosimann for preparing the ion transporter sequences. This work was supported by a Swiss National Science Foundation research professorship grant to P.M. and the Roche foundation.

References

[1] S. R. Eddy, "Hidden markov models.," *Curr Opin Struct Biol*, vol. 6, pp. 361–365, June 1996.

[2] M. Dean and T. Annilo, "Evolution of the atp-binding cassette (abc) transporter superfamily in vertebrates.," *Annu Rev Genomics Hum Genet*, vol. 6, pp. 123–142, 2005.

[3] R. D. Finn, J. Mistry, B. Schuster-Bockler, S. Griffiths-Jones, V. Hollich, T. Lassmann, S. Moxon, M. Marshall, A. Khanna, R. Durbin, S. R. Eddy, E. L. L. Sonnhammer, and A. Bateman, "Pfam: clans, web tools and services," *Nucl. Acids Res.*, vol. 34, no. suppl 1, pp. D247–251, 2006.

[4] S. R. Eddy, "Profile hidden markov models.," *Bioinformatics*, vol. 14, no. 9, pp. 755–763, 1998.

[5] R. Chenna, H. Sugawara, T. Koike, R. Lopez, T. J. Gibson, D. G. Higgins, and J. D. Thompson, "Multiple sequence alignment with the clustal series of programs.," *Nucleic Acids Res*, vol. 31, pp. 3497–3500, July 2003.

[6] S. B. Needleman and C. D. Wunsch, "A general method applicable to the search for similarities in the amino acid sequence of two proteins.," *J Mol Biol*, vol. 48, pp. 443–453, March 1970.

[7] E. Jones, T. Oliphant, P. Peterson, *et al.*, "SciPy: Open source scientific tools for Python," 2001–.

[8] T. Kohonen, *Self-Organizing Maps*, vol. 30 of *Series in Information Sciences*. Berlin: Springer, 3 ed., 2001.

[9] Hoekstra, Kraaijveld, de Ridder, and Schmidt, "The complete sprlib and annlib," 1996.

[10] N. Fankhauser and P. Maser, "Identification of gpi anchor attachment signals by a kohonen self-organizing map," *Bioinformatics*, vol. 21, pp. 1846–1852, May 2005.

[11] D. M. Beazley, "Automated scientific software scripting with swig," *Future Gener. Comput. Syst.*, vol. 19, pp. 599–609, July 2003.

[12] R. Ihaka and R. Gentleman, "R: A language for data analysis and graphics," *Journal of Computational and Graphical Statistics*, vol. 5, no. 3, pp. 299–314, 1996.

[13] M. B. Eisen, P. T. Spellman, P. O. Browndagger, and D. Botstein, "Cluster analysis and display of genome-wide expression patterns," *PNAS*, vol. 95, no. 25, pp. 14863–14868, 2004.

[14] A. J. Saldanha, "Java treeview, extensible visualization of microarray data," *Bioinformatics*, vol. 20, no. 17, 2004.

[15] L. Kaell, A. Krogh, and E. L. L. Sonnhammer, "A combined transmembrane topology and signal peptide prediction method," *Journal of Molecular Biology*, vol. 338, no. 5, pp. 1027–1036, 2004.

[16] J. Ward, "Identication of novel families of membrane proteins," *Bioinformatics*, vol. 17, p. 560563, 2001.

[17] N. Brown, L. C., and S. C., "Mview: A web compatible database search or multiple alignment viewer," *Bioinformatics*, vol. 14, no. 4, pp. 380–381, 1998.

[18] J. Soeding, "Protein homology detection by hmmhmm comparison," *Bioinformatics*, vol. 21, no. 7, pp. 951–960, 2005.

[19] P. Paindavoine, S. Rolin, S. Van Assel, M. Geuskens, J. C. Jauniaux, C. Dinsart, G. Huet, and E. Pays, "A gene from the variant surface glycoprotein expression site encodes one of several transmembrane adenylate cyclases located on the flagellum of trypanosoma brucei.," *Mol Cell Biol.*, vol. 12, no. 3, p. 12181225, 1992.

Chapter 6

List of Programs

1 GPI-SOM

Identification of GPI-anchor signals by a Kohonen Self Organizing Map (SOM).

- Input: Multiple protein sequences in FASTA format
- Output: Position on the SOM for each sequence as well as the predictions whether the GPI signal sequence has been found, not found or is undecidable. See Figure 1.
- Programming language: C
- Web: http://gpi.unibe.ch
- Source: http://gpi.unibe.ch/kohgpi-1.5.tar.gz

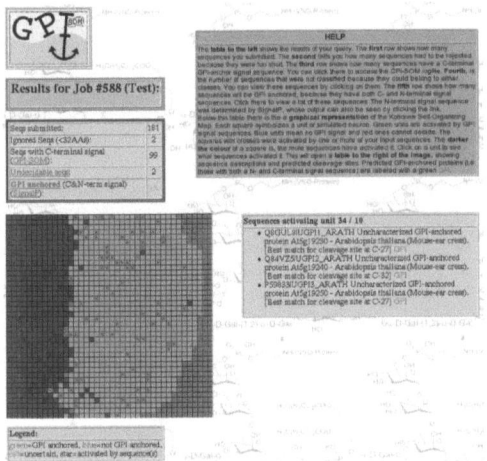

Figure 1: Screenshot of the result-page of the GPI-SOM Web-Interface.

2 Reptile

Finds all repeats in proteins using a word counting method combined with probability based sorting mechanism.

- Input: Multiple protein sequences in FASTA format.
- Output: List of perfect (but not necessarily direct) repeats found in each protein, sorted by P-value. See Figure 2.
- Programming language: C++
- Web: http://reptile.unibe.ch
- Source: http://genomics.unibe.ch/software/reptile.tar.gz

```
> reptile test.fasta
Maximal word size: 20
p-value threshold: 0.0001
>Tb11.0330
19x AAVPQPPM (p=1.50079e-154)
9x AAVPQPPMAAVPQPPM (p=8.07233e-149)
7x PQPPMAAVPQPPMAAVPQPP (p=7.23037e-142)
5x MAAVPQPPMAAVPQPPMAAV (p=2.28646e-93)
>Tb10.6k15.0030
26x EP (p=1.30273e-41) => DIRECT <=
>Tb927.7.3440
5x AEAEARAR (p=7.73643e-33)
3x AEAEARARAEAEA (p=2.26724e-28)
2x AEAEARARAEAEARAR (p=3.54309e-17) => DIRECT <=
>Tb927.8.8160
2x ESRKPSVKDLKRPPLHGAPS (p=2.9197e-21)
```

Figure 2: Example of using Reptile on some highly repetitive proteins from *T. brucei*.

3 Dora

Database of repetitive antigens.

- Authors: Tien Nguyen-Ha, modified by Niklaus Fankhauser
- Output: Proteins from all organisms in the integr8 database can by displayed by the p-value of the repeats they contain and whether they have GPI-anchor, transmembrane domains or N-glycosylations. See Figure 3.
- Programming language: PHP
- Web: http://genomics.unibe.ch/dora
- Source: http://genomics.unibe.ch/software/dora.tar.gz

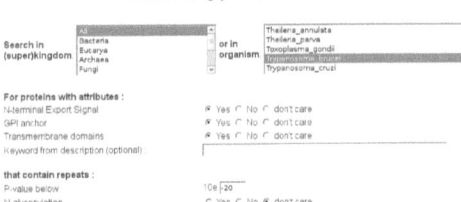

Figure 3: Screenshot of the database query page from Dora.

4 Org-Get

Performs automatic updates of the Dora database whenever the integr8 database is changed.

- Input: Fetches the index of the integr8 database server by FTP every day to check if any proteome has been added or changed.

- Output: Processes each new or changed proteome by GPI-SOM, Reptile, Phobius and a program searching for N-glycosylation patterns. Stores results in the Dora MySQL database. See Figure 4.

- Programming language: Python

- Source: http://genomics.unibe.ch/software/OrgGet.tar.gz

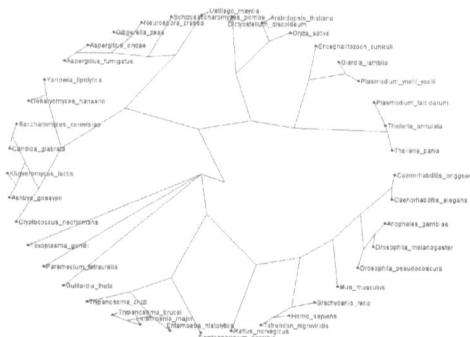

```
Downloading report file from ftp.ebi.ac.uk
UNCHAN> Aeropyrum pernix, 1702 Proteins on FTP,
1702 Proteins in MySQL
REJE> Campylobacter jejuni, 1623 Proteins,
Bacteria
NEW> Leishmania infantum, 7871 Proteins, Eucarya
New: 1
Changed: 9
Rejected: 246
Redundant: 145
Unchanged: 199
ToDo: 10
Org-Get Update Start mail sent!
Work directory exists -> Emptied.
Processing 1 of 10:
Desulfovibrio vulgaris (3085 prot)
Downloading proteome...
Transfering sequences and descriptions to SQL...
Running gpisom...
```

Figure 4: Example of a run of Org-Get. Similar lines of output were removed. Org-Get is configured to reject new proteomes unless they are eukaryotic.

5 circDend

Draw a circular dendrogram from Cluster 3.0 output. Cluster was written to cluster microarray data, but can be used for clustering almost anything. This program visualizes the clustered data as a circular dendrogram, similar to what Java TreeView does in a linear way.

- Input: CDT and GTR files created by Cluster 3.0 from multidimensional data.

- Output: Graphical rendering of the tree. See Figure 5.

- Programming language: Python

- Source: http://genomics.unibe.ch/software/circDend.py

Figure 5: Plot of a phylogenetic tree created by circDend. Clustered data: Sugar, Amino Acid and Nucleoside/-tide transporter Pfam hit numbers in eukaryotes.

6 memProtPlot

Create plots of transmembrane proteins, based on the Phobius or HMMTOP prediction. Amino acids can be colored by 60 selectable presets.

- Input: One protein sequence containing transmembrane domains in FASTA format.

- Output: Graphical rendering of the amino acid sequence along the predicted membrane topology of a transmembrane protein. See Figure 6.

- Programming language: Python

- Web: http://genomics.unibe.ch/memplot.html

- Source: http://genomics.unibe.ch/software/memplot.tar.gz

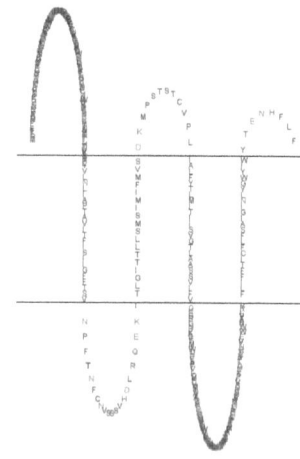

Figure 6: Plot of the sequence on the Phobius topology prediction of an acetylcholine receptor by memProtPlot.

7 ReduceToMax

ReduceToMax trims a set of proteins until there is no pair of sequences left with a similarity above a given threshold. This is useful to minimize redundancy in a list of proteins in order to avoid bias of input sets to build a profile. All sequences are aligned with all sequences. If two sequences are similar above threshold, the shorter is removed. The result page shows what and how many sequences have been removed. It is possible to automatically build a HMM profile from the reduced set of proteins and search for matches in the organisms in integr8.

- Input: Multiple protein sequences in FASTA format.

- Output: Protein sequences in FASTA format with redundancy and bias removed. See Figure 7.

- Programming language: Perl, alignments done with an in-line C function.

- Source: http://genomics.unibe.ch/software/rtm.tar.gz

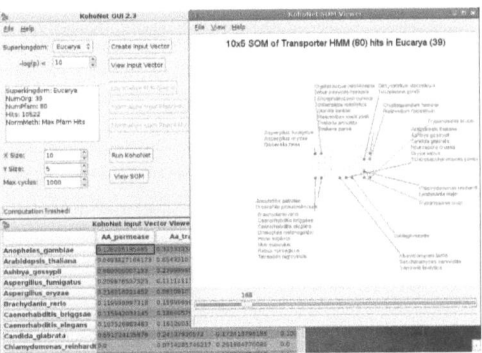

Figure 8: Screenshot of KohoNet-GUI in use to find patterns in eukaryotic organisms based on a transporter profile search.

```
Threshold: 50%
entdat/job9.fasta read: 11 seqs
Q4Q6D5 - Q4Q6D4: 65.66% similar-> Q4Q6D5 removed
Q4Q6D6 - Q4Q6D4: 52.68% similar-> Q4Q6D4 removed
Q4QBF4 - Q4QBE9: 50.32% similar-> Q4QBE9 removed
Q4QBF4 - Q4QBF3: 72.45% similar-> Q4QBF3 removed
4 similar sequences removed
entdat/job9.txt written: 7 seqs
```

Figure 7: Example of a run of ReduceToMax on a set of MRP proteins.

8 KohoNet-GUI

A Kohonen neural network in C with GUI and control in python to evaluate organism transporter profiles.

- Input: MySQL table of HMM profile search results.

- Output: Self organizing map trained to find groups in the input pattern. See Figure 8.

- Options: Three different normalisation methods, adjustable p-value threshold for the inclusion of profile search hits, SOM size and the number of training cycles.

- Programming language: Python and a C function used as a module with SWIG.

- Source: http://genomics.unibe.ch/software/kohonet.tar.gz

9 PfamOrg

Creates overview of graphs for pfam transporter profiles of all organism of a superkingdom as well as detailed information for certain Pfam hits in any organism.

- Input: MySQL table of HMM profile search results.

- Output: A graph for each organism, where the X-axis represents the numbered Pfam profiles and Y-axis is the number of hits for this profile in an organism. See Figure 9.

- Options: By positioning the mouse over a bar it is possible to see which Pfam profile it represents in the Web-Interface. Clicking on the bar displays all hits for a Pfam profile in a certain organism. Each superkingdoms of organisms can be selected for an overview in the Web-Interface.

- Programming language: Python

- Web: http://genomics.unibe.ch/PfamOrg.html

- Source: http://genomics.unibe.ch/software/PfamOrg.tar.gz

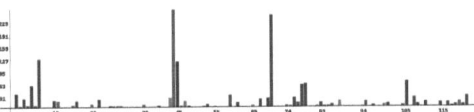

Figure 9: Pfam transporter profile search results in *H. sapiens* plotted by PfamOrg.

10 paraFree

Box plot analysis of differences in parasites and free-living organism protein type percentages.

- Input: MySQL table of HMM profile search results.

- Output: Box plot of the percentages of proteins found by a profile search in parasites and free-living organisms. A Mann-Whitney U test is performed to estimate the statistical significance. See Figure 10.

- Programming language: Python, using Ploticus for graphs

- Source: http://genomics.unibe.ch/software/paraFree.tar.gz

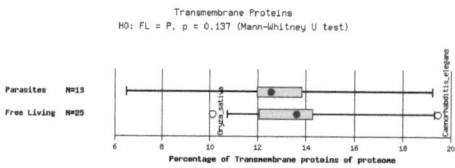

Figure 10: Data created by paraFree plotted by Ploticus.

11 protCluster

Creates Protein-Families of a transmembrane-subset of a set of organism by statistically guided clustering of pairwise alignment score-matrices. Alignments were done using a C implementation of the Needleman-Wunsch algorithm. Clustering parameters were determined automatically by a program written in python which maximizes desired family shapes. Family Hidden-Markow-Models were searched against those of all organism. Results were then analyzed, annotated, visualized and clustered.

- Input: Proteome files in FASTA format.
- Output: MySQL table of clustered protein families and results of the HMM-HMM searches. See Figure 13 for an example protein family and Figure 12 for the complete pipeline.
- Programming language: Python
- Source: http://genomics.unibe.ch/software/protClust.tar.gz

Figure 11: Multiple alignment of a family of Nucleobase transporters created by protCluster, visualized by TransOrgalin.

12 TransOrgalin

A web application designed to visualize protein families and their similarities in different organisms. It is an interface to the data created by protCluster.

- Input: MySQL tables created by protCluster
- Output: List of families, details about families and family similarities with scaled graphical display of a multiple alignment, occurrences of families in different organisms, recursively searched relations of families. See Figure 13.
- Programming language: Python
- Web: http://transorgalin.unibe.ch
- Source: http://genomics.unibe.ch/software/transorg.tar.gz

TransOrgalin is composed of eight modules, which are described here:

```
1. dpp_dist
   Requires: dpp
   Input:  org_sequence.txt
   Output: org_matrix.txt

2. protClust-Num.py
   Requires: Python statistics module, NumPy
   Input:  org_matrix.txt
   Output: org.cNtNsN.pickle

3. pickle2sql.py
   Requires: SQLite, clustalw, mview, html2png.py,
             newicktops, convert, mogrify,
             hmmbuild, hhsearch, hhmake
   Input:  org.cNtNsN.pickle
   Output: org.cNtNsN.d/, SQL: families,
             cluster_params, metafam (famdesc)

4. N times hmm-hmm.sh
   Requires: respars.py, SQLite
   Input:  org.cNtNsN.d
   Output: HHR-org.cNtNsN.d-VS-org2.cNtNsN.d
```

Figure 12: The pipeline used to compute all HMM-HMM search results on the University of Berne Linux Cluster (ubelix). Dpp_dist is a Perl program that splits the computation of the alignment score matrix into equal part to be computed in parallel on ubelix. Each part will be computed using the same C implementation of the Needleman-Wunsch algorithm used in ReduceToMax, which is called dpp (dynamic programming pathfinder). The proteins are then clustered by protClust-Num.py, using the NumPy and statistics module. Clustalw is used to create alignments of these families. Mview creates an HTML visualisation, which is converted to a scaled image by html2png.py. Transmembrane domains are also plotted in the same scale by this program using the Dora MySQL database. Trees are created by newicktops (NJplot) and converted to PNG using ImageMagick. Hmmer builds profile HMMs of the families, which are searched against all other HMM profiles using HHsearch. SQLite is used to store the results on ubelix, because MySQL is not available.

12.1 FamiList

A table to lists protein family names of specified organism which were derived from annotations of the family members. The number of members and average TM domain and protein length for each family can be seen. The next column shows similar protein families found by hmm-hmm search from other organism. Raising the pev-threshold (Family VS all organisms) protein hmmsearch), activates the next column to display similar proteins of other organism. The list of family names is searchable. For example, "abc" could be searched to find all protein families containing "abc" in their annotation. A p-value threshold for displaying hmm-hmm results can also be specified. It is possible to apply filters: "unique" mean families with no hmm-hmm results in other organism, "mutual" the opposite, "exciting" families which are unannotated but have hmm-hmm results, "parasite" are families which have no hmm-hmm results from non-parasites, "kinetoplastidae" are families which only have similarities to families of that taxonomic group and "nonhuman" are families with no similarity to human families. Below the list are the percentages of all similar proteins to the above protein families for each organism. Click on the families gets the info-page about them containing dendrogram, graphical or HTML alignment, positions of TM domains in relative to the alignment and protein annotations and sequences. Fam-

Figure 13: The family occurrence display (FamiOccurence) of TransOrgalin.

ily similarities are also clickable to view the info-page of the unified family-family dendrogram and alignment.

12.2 FamView

To the left is the dendrogram of the protein sequences of this family. Below that is a graphical projection of the aligned protein sequences in this family, which is scaled to a constant width and coloured according to similarity and amino acid property. Underneath are the positions of transmembrane domains in the alignment at the same scale. The last item on this page is a list of similar proteins families found by HHsearch below the selected threshold. The names of the similar protein families are clickable to get to a page where an analysis of the pool of proteins of the joined families can be seen.

12.3 Statistics

A table unifying the information of the FamiLists of all organism. For every organism, it shows the number of families, clustering parameters, total proteins, number of transmembrane proteins, number of proteins in families, average family size, and the number of results of applying all the filters mentioned in FamiList (exciting,unique,parasite,nonhuman). The whole table is influenced by the selected p-value. Rebuilding the statistics for an new p-value threshold takes some seconds unless it's already cached. An additional column called vExcit, which shows the number of protein families which have no or only hypothetical annotations in one organism, but good annotations in similar families found by HMM-HMM search in other organism, and yet show no such annotation by searching the family HMMs against a database of all proteins used in this study. This number is influenced by the p-value threshold selected.

12.4 FamiOccurence

Tabular representation (Figure 13 of the occurrence of protein families in all organism. Sortable by organism in adjustable priorities. The protein family clusters are reduced in the organism of sort priority one, so all other organism are relative to this one. The family id-numbers are clickable and lead to the FamView of the selected family. This meta-clustering is influenced by the similarity selected p-value threshold. The family name in the first column in determined from a non-hypothetical family annotation in that row.

12.5 FamiRelations

Starting from one family, this searches iteratively as long as it takes (again with the results) to find all connected protein families above the selected threshold. The first row (labeled in the first column as level 1) contains the original protein family from where the search starts. The second row contains all protein family found by searching with the original protein. The third row all families found by searching with the families of the second row, and so on. Until no more protein family is found that has not been found before on a lower level. Each table cell contains first the organism-family-Id (OxxFxx) of the search query family and second the Id of the result family followed by the family name and organism abbreviation. The family name is clickable and leads to FamView. Below the table is the number of families that were found to be related on some level and the total number of families in this study.

12.6 ClusterSearch

Starting from a keyword search in all family descriptions, cluster the results of searching with the results above a certain threshold. The search works iteratively like in the FamiRelations display. View dendrograms as vertical or circular trees or with optional symbols for better overview. An example search keyword can be selected from the list to the right. Clustering is performed by Cluster 3.0 with the pairwise complete-linkage method using euclidean distances. The resulting tree-file is visualized using the python imaging library and made interactive by XHMTL image-maps.

12.7 ClusterCompare

Show family relations between a selectable group of organism above a certain threshold. Again in vertical, graphical or circular rendering of the tree, as well as with the ability to cut off the top nodes of the tree in order to display a number of sub-trees, which are often meaningful units, such as proteins families with similar function in different organism of which some could even lack annotation in some of them. The clustering options and methods are the same as in ClusterSearch.

12.8 PoolView

The analysis of the pool of a pair of similar protein families. First a table comparing description, of number of members, average length, and average number of transmembrane domains for the similar families of both organism. There is dendrogram of the pooled sequences of both families.

Chapter 7

Abbreviations

ABC	ATP binding cassette
ATP	Adenosine triphosphate
BLAST	Basic Local Alignment Search Tool
BLOSUM	BLOck SUbstitution Matrix
cDNA	Complementary (to mRNA) deoxyribonucleic acid
COS	Cercopithecus aethiops
CRAM	Cysteine-rich acidic membrane protein
CSP	Circumsporozoite protein
DNA	Deoxyribonucleic acid
DORA	Database of repetitive antigens
GBP	Gycophorin-binding proteins
GPI	Glycophosphatidylinositol
GPI2	Phosphatidylinositol N-acetylglucosaminyltransferase
EMP	Erythrocyte membrane protein
EST	Expressed sequence tag
HMM	Hidden Markov model
ISG	Invariant surface glycoproteins
MAEBL	Apical membrane erythrocyte binding antigen
MALDI-TOF	Matrix-assisted laser desorption/ionization
MD	Megadalton
MESA	Mature parasite-infected erythrocyte surface antigen
mRNA	Messenger ribonucleic acid
MS	Mass Spectrometry
MSP	Merozoite surface protein
PCA	Principal components analysis
PCR	Polymerase chain reaction
PGRS	Polymorphic GC-rich repetitive sequence
RESA	Ring-infected erythrocyte surface antigen
RNA	Ribonucleic acid
rRNA	Ribosomal ribonucleic acid
SOM	Self-organizing map
TransOrgalin	Transmembrane families in multiple organisms using alignments
TRIPS	Tandem Repeats In Protein Sequences
UPGMA	Unweighted Pair Group Method with Arithmetic Mean
VSG	Variable surface glycoproteins

Die VDM Verlagsservicegesellschaft sucht für wissenschaftliche Verlage abgeschlossene und herausragende

Dissertationen, Habilitationen, Diplomarbeiten, Master Theses, Magisterarbeiten usw.

für die kostenlose Publikation als Fachbuch.

Sie verfügen über eine Arbeit, die hohen inhaltlichen und formalen Ansprüchen genügt, und haben Interesse an einer honorarvergüteten Publikation?

Dann senden Sie bitte erste Informationen über sich und Ihre Arbeit per Email an *info@vdm-vsg.de*.

Sie erhalten kurzfristig unser Feedback!

VDM Verlagsservicegesellschaft mbH
Dudweiler Landstr. 99　　　　　　　Telefon　+49 681 3720 174
D - 66123 Saarbrücken　　　　　　　Fax　　　+49 681 3720 1749
www.vdm-vsg.de

Die VDM Verlagsservicegesellschaft mbH vertritt

Printed by Books on Demand GmbH, Norderstedt / Germany